Courtesy Glenn L. Martin Co.

THE "BALTIMORE" BOMBER

THE AIRCRAFT APPRENTICE

BY

LESLIE MACGREGOR

Member: American Society of Tool Engineers

Originally published in 1942

ISBN: 978-1-940001-33-3

The Aviation Collection
by
Sportsman's Vintage Press
2015

FOREWORD

The aeronautical industry has grown rapidly during the past decade, but during the present national emergency it has reached an all-time high level of production. The building of modern airplanes requires well-trained and highly efficient skilled and semiskilled workers.

Boys and girls in the public schools are intensely interested in the airplane and its performance. The construction of airplanes has appeal to most boys. They are hungry for reliable and authentic information. This book, *The Aircraft Apprentice,* will fill a great need.

To the youth looking forward to employment in the airplane industry, *The Aircraft Apprentice* will give a better understanding of the workers and their jobs. It is written in plain language, easily understood by boys of the vocational, junior and senior high school ages. The information presented will be helpful in assisting students to make their choice of the aviation occupations in line with their aptitudes and abilities.

This book will also be a great aid to boys who are in training for the various trades in the airplane industry. They will be able to secure firsthand a reliable knowledge of the materials, machines, tools and the various processes used in the construction of airplanes. It will be the companion not only of men in school, but of those entering upon work as apprentices in this great industry.

The Aircraft Apprentice will be of value in both public and school libraries of America. It can be used as a reference wherever students are investigating the possibilities of preparing for employment as aviation apprentices. Under certain conditions, it will serve as a text.

CHARLES W. SYLVESTER
Director, Division of Vocational Education Baltimore, Maryland

To a heroic airman
of the Royal Air Force,
Ivor Gwessyn Thomas,
this book is dedicated.

There are things to conquer, not men;
there are fields to be won, not of soil
but of understanding.
Go, gird thyself, O Youth,
with knowledge.

Leslie MacGregor.

AUTHOR'S PREFACE

During the past few years a definite need has arisen for more men who possess mechanical skill. It is indeed strange that a shortage should exist in this particular field, when one considers that nine boys out of every ten show definite mechanical instincts very early in life.

No doubt a large percentage of parents have stifled these natural instincts and have advised their sons that the desk is more conducive to refinement and position than the workbench. Thus, the so-called white-collar job has lured many thousands of young men into positions for which they do not have the proper enthusiasm. The resulting overcrowding has retarded their promotion.

The Machine Age is here, but it will surely melt away, as did the Ice Age, if we do not encourage the mechanical instincts of our boys. We must train them in the shops so that these instincts will be converted in practical skill.

Therefore, the purpose of this book is to secure the interest of those boys who possess a mechanical instinct yet know nothing as to how it is practiced and are desirous of entering that constantly expanding field, the manufacture of aircraft.

LESLIE MACGREGOR

CONTENTS

CHAPTER 1

AIRCRAFT MANUFACTURING AS A CAREER

Choosing a Vocation—How important it is that the boy who is nearing the end of grade school should start forming some definite opinions as to what place he is going to fill in industry. Soon he will be entering high school, and if he knows what he wants, he can receive some very valuable training in his chosen field. But unlike the four-year-old, who makes the snap decision that he wants to be a fire fighter or a railroad engineer, the older boy is faced with a very difficult problem.

How is he to realize the various phases of industry, or what phase or what industry will suit him best? During his school life, the average boy is more interested in sport than in wrestling with industrial problems—and rightly so. In this way, his physical standard is raised and his sense of fair play is developed, but still he is in the dark as to what happens behind factory walls.

In this very important matter of choosing a vocation, the boy needs proper guidance from three separate sources: first, from his schoolmaster, who understands more than anyone else the boy's capabilities; then, from his parents because he is still a great responsibility to them; and finally, from his own inclinations. A judicious survey of these three factors has every possibility of placing in industry an enthusiastic apprentice. It is from such beginnings that factory executives grow.

The following pages are intended to help the boy himself, in forming his own opinions as to a possible career, by pointing out the various phases of the manufacture of aircraft.

So, therefore, it is well that he should start thinking about the future. Is it not possible that countries, now widely separated by distance and ideals, may in the future be brought to a closer understanding by aerial travel? Is it farfetched to visualize the swift exchange of national products by aerial freight car, or people spending short vacations on remote tropical islands?

Surely, one should not be termed a dreamer for exploring the possibilities of a future career among the highly skilled branches of the aircraft industry. But let us also consider its glorious past history and how the airplane rapidly evolved from a tricky experiment to an approved mode of traveling.

History of Aircraft—At the present stage in the development of the airplane, the greatest laboratory of all is still that elemental airy space where an aircraft may fly. Steels, alloys and structures are developed in the earth-bound laboratories and workshops; engines are calibrated on the test benches; and pilots have to be approved by the medical profession. But it is only in full flight that the working combination of all these can be appraised.

The science of flight, therefore, was not developed entirely through the quiet pursuit of academic studies. Ever since the first pilot left the ground, knowledge has thundered above the roar of every take-off, has turned the rhythmic drone of every flight into an object lesson in which air current, stability, power, and human reactions have been analyzed.

After the close of the Great War of 1914-18, the business of building airplanes was at a very low ebb. The demand for commercial aircraft was nonexistent, for the simple reason that at that period airplanes were far too dangerous for public use, and the powers which had been locked together in ruthless brutality for four years were so impoverished that aerial expansion was impossible.

Barnstormers—The flying was done for the most part by

young daredevils, called "barnstormers," who flew about the country in war-time equipment, taking people for five-minute rides at two or three dollars per passenger. However, the financial returns were such that these daring young pilots would ofttimes find it necessary to catch up on their sleep within the drafty confines of the nearest barn (hence the name barnstormers) with or without the consent of the farmer. Sometimes they would tuck themselves into their open cockpit, as if to fly away into the arms of Morpheus.

It was a hard and dangerous life, with success or disaster depending on the tautness of their flying wires, the soundness of their wooden spars, and the clarity of their own judgment in an emergency. But it was these young enthusiasts who finally made America "air conscious." It was they who gave Mr. and Mrs. America their first flip in a flying machine.

Although the first actual power-driven flight was accomplished in December 1903, this was the climax of an ambition that had lingered in the hearts of men for centuries. Why then, you may ask, did it take so long for men actually to fly? In a nutshell the answer is: They had to wait until they had developed a suitable engine.

Famous Flights—The Wright brothers, Wilbur and Orville, were the first ones to apply successfully motive power to the gliding machines, which they and others had been experimenting with for years. With tremendous patience and enthusiasm, these two Americans labored in obscurity and often in the face of ridicule, until in 1903, their machine made a hop of 0.98 of a mile and stayed in the air for 59 seconds. During these fateful moments, a great industry was born, and another element was made available for man to explore.

The two brothers, however, did not proclaim their triumph to the world as would have been expected. Instead, they went back to their little workshop to incorporate their newly found knowledge in their flying machine. One year later, they made another flight, this time covering a distance of three miles, and staying in the air for a period of five minutes, twenty-seven seconds.

After another year of experimenting, they were able to fly twenty-four miles, and remain in the air more than half an hour.

The rest of the world then caught the craze. In the year 1909, the English Channel was crossed by air by M. Louis Bleriot, an intrepid French aeronaut, and so year by year, an excited world has witnessed one success after another in the development of the airplane.

In tribute to the ingenuity of man, consider the courageous flight of Captain Alcock and Lieutenant Brown, of the British Royal Air Force, taking place a mere ten years after the first flight across the English Channel. This time, the broad Atlantic was spanned nonstop by a heavier-than-air machine for the first time. Using a Vickers Vimy machine that was built as a bombing plane, the two airmen took off from a clearing above Trepassey Bay in Newfoundland and landed at Clifden in Ireland. How true it is that "Peace hath her Victories no less renowned than War."

Having dropped their landing gear shortly after the take-off for lightness, the plane skidded in for a landing on its belly. Another chapter in the history of the airplane had thus been completed.

The flight that did more than any other to commercialize the use of the airplane was the historic flight by Charles A. Lindbergh from New York to Paris in 1927. The population of both the old and the new worlds were already conscious of the possibilities of the airplane as a method of transportation, and Lindbergh's flight demonstrated this so conclusively that the aircraft industry received an impetus that has never abated.

In this world there are still a few inaccessible places that stand as a challenge to the adventurous spirit of mankind.

Until a few years ago, Mount Everest, the highest mountain in the world, which is estimated to rear itself 29,000 feet into space, was one of those places that defied exploration. Many expeditions had set out to reach this roof of the world, but all had ended in failure, some with disaster. It was not until 1933 that the noble crest of Mount Everest was finally exposed to the prying eyes of man.

In that year, a flight of British airmen flew to far-off India, set up a base near the Himalayas, and made two successful flights over the summit of this "ageless giant." It is not for my inadequate pen to try to describe the magnificent effort of these men, but I gladly refer you to that engrossing tale, *Pilots' Book of Everest* by Clydesdale and M'Intyre.

During the first phase of the history of the airplane, which ranged from its inception until just after the first World War, the industry was more concerned with problems of design and performance than with manufacture. Each of the pioneers was a man who seemed to be able to combine the profound knowledge of the engineer, the vision of the designer, the skill of the mechanic and the daring of the pilot.

After the fundamental laws of flight were fairly well established, the pioneers were retained as aeronautical engineers to work on design only, and new blood was called upon—including tool engineers, production engineers, and more skilled workmen in the various mechanical branches of the industry.

Although in most major industries today the accepted practice is based on the accumulated knowledge of many years, these specialists soon found out that the problems of building aircraft were entirely new. They had to develop new methods and processes in order to manipulate the new materials which were making their appearance. Before the advent of the airplane, mechanical products were usually built to specifications for strength rather than lightness—the latter, in more cases, being sacrificed for the former. In the aircraft industry, those two age-old enemies are reconciled, and engineers are continually working and studying to combine them more efficiently.

The Industry of Youth—It is because of such new problems that the aircraft industry is one in which youth predominates. New problems require the attention of brains that are unfettered by old ideas. Not only must the aircraft manufacturer create a structure that is phenomenally light for its strength, with materials that have never been used in any other industry, but he must develop high-production methods in order to produce sufficient quantities to meet the tremendous military needs of the nation. The higher the production, the higher the income, making possible the tremendous expenditure for experimentation. So highly specialized have the different branches of this industry become that it is possible to carve out a career in any one of them.

Inspiration and Enthusiasm—The sight of an airplane winging its way along the aerial highway always seems to inspire the average schoolboy with its challenge of new worlds to conquer. He catches some of that enthusiasm that has always motivated the craftsmen and technicians of the entire engineering field. Like the mainspring of a watch, enthusiasm stores up untiring energy and releases it to the one who has the will to use it. Without this human capacity for utilizing enthusiasm, the world would be a dismal place indeed. Columbus himself would have lived and died in obscurity. Billions of horsepower would have been wasted in the condensation of unharnessed steam, had not James Watt caught that coil of enthusiasm that enabled him to turn the wheels of industry.

Importance of Being Properly Placed—It is very important, however, that the youth who is thus enthused with the romance which surrounds the completed airplane should also realize the possibilities for him in the manufacturing.

Nor should his enthusiasm blind his perception of what phase of the industry he should enter. This question must be studied very carefully. If he selects a job for which he is not suited, he becomes what is known as a misfit. In this condition he causes trouble. He loses interest in his job and his workmanship becomes inferior. Soon his dreams of a career go into a tailspin, and he cannot pull out unless the personnel department gets hold of the controls and brings him to a happy landing in a department more suited to his inclinations and mechanical ability.

Boyish Characteristics to Suit Jobs—It is one of the most difficult of tasks to try to influence any boy in making his initial choice of a job. The boy must choose. We approve, advise, or call for another choice; we do not dictate. To do so would be to crash the walls of his self-reliance, something he will need later in life. In this advisory capacity, therefore, it is suggested that the boy who comes through with flying colors in his academic subjects, who is studious to a degree especially in mathematics, who has vision in a practical way and a creative mind, may well look towards the aeronautical engineering department as the scene of his life's work. On the other hand, the boy who craves action, who likes not only to "take things apart" but also to put them together again, who is laboriously patient and has constructive ideas, who is practical, rugged and unafraid of hard work, is the one who should actually build the aircraft in the shop.

The particular shop department for which he is fitted depends on other, less fundamental, but none the less important, whims. For instance, the habitual wood whittler had better make for the wood shop. The one who can find within himself a natural bent for detailed accuracy, a glutton for study, thriving on mechanical problems, should head for the toolroom. The hard worker, who is not so much interested in doing a variety of jobs but who wants to become a specialist in repetitive work, may find his place in the machine shop, sheet-metal machines, small-parts assembly, hull or fuselage assembly. Those who want to learn the maintenance and servicing of aircraft should be destined for the final assembly. The fellow who understands and likes geometry, and who possesses above all a pair of tricky, ingenious hands, should enter that department of artists called "sheet-metal workers."

Individual success in any one of these departments depends entirely on the amount and quality of the individual effort.

CHAPTER 2

VOCATIONAL TRAINING

Early Training—Anything that is moving towards a goal needs guidance. It would also seem that the more desirable the goal, the more difficult is the pathway. In the attainment of his goal, the apprentice must be guided by good sound training.

But he must realize that the acquisition of this training in itself is no easy matter. Actually, the training period extends long after he starts to work, so he must be prepared both to work and to study. The boy most likely to succeed is he who realizes that he must acquire a knowledge of the theoretical as well as the practical, and who learns to combine them to the best advantage.

As the fencing expert knows when to parry, when to thrust, so must the apprentice learn the skillful application of theory and practice. As a knowledge of the theoretical is gained mostly in the classroom, it is important that advantage of this should be taken in high school, when the mind is most pliable. He who waits until he starts to work before realizing the link between the classroom and the shop has lost at least four valuable years. Furthermore, in some cases, he stands a chance of not getting into his chosen field, unless he has had previous training in the classroom.

What to Study—The boy in high school who wants to get ahead in the aircraft industry should concentrate on plane geometry, solid geometry, mensuration, algebra, elementary trigonometry and engineering drawing.

A knowledge of the geometrical subjects is especially valuable in layout work. Any flat layout is merely one or a combination of plane geometric figures. The principles of solid geometry are followed when laying out any structure that has the three dimensions of length, breadth and thickness. A knowledge of mathematical subjects is important, as it is very often necessary to calculate unknown dimensions and angles from given dimensions. An understanding of the principles of engineering drawing is also necessary to the success of the aircraft apprentice. The ideas of the engineer are carried to the shop in the form of drawings, and by interpreting these correctly, the mechanic builds in metal.

By taking full advantage of this type of training, the apprentice may influence for the better his whole life's work. Indeed, he should consider his apprenticeship as beginning in high school, so that when he finally enters the factory, he can impress the foreman with his technical accomplishments.

Remembering the necessary co-ordination of the theoretical and the practical, the prospective apprentice asks now this logical question: "How can I supplement my high school technical training with practical training, before I enter the factory?"

Aid from the Personnel Director—In order to explore all the possibilities for training in his particular neighborhood, the prospective apprentice should get in touch with the Personnel Director of the nearest aircraft factory. In most cases this official is a man of keen psychological insight, and is in touch with all the training institutions within a large radius.

A. Vocational Training—In America the vocational training school system has advanced to a very efficient stage. The development of high-production methods forced out the old apprenticeship system, whereby the apprentice received from the employer his lodging, food, clothing and complete training. In the factory, whose output is on a high-production basis, there is no room for the young man who has no knowledge at all of the mechanical practices involved. He must enter the factory with the idea of going to work to produce.

In recent years the demand for the production of aircraft has been so great that manufacturers all over the country have been co-operating with the different local education authorities to prepare courses in certain phases of practical aircraft work. The manufacturers have even gone so far as to supply some of their skilled men to work as teachers. This has proved to be an ideal arrangement, with the result that thousands of young men have been started in various aircraft factories with a confidence and skill that would have taken many months to acquire in the shop.

B. National Defense Training—The demands upon the industry due to national defense have brought about the adoption of the short, intensive vocational course, usually of about eight weeks, five days per week and eight hours per day in duration. Although it is not expected that the boy should become a fully skilled workman in this short period, he is of immediate value to his foreman because he can lay out the spacing of a row of rivets, drill straight clean holes of the proper diameter, apply a section of "skin," lay it out and trim it. He understands the heat-treatment and aging of rivets and the necessity of care in handling aluminum and its alloys. He has already made riveted assemblies, used angle drills for inaccessible places, and removed bad rivets. He has even developed and formed a fuselage frame and riveted it in place. This type of training is fundamental in character; there is no time for superfluous instruction. Electric- and acetylene-welding courses are also offered, as well as machine-tool operation.

Some of these schools maintain a battery of punch presses, the dies being loaned by certain aircraft factories. Under the guidance of expert instructors, the students are taught the secrets of "die set-up," so as to produce formed pieces that comply with the specifications.

With each stroke of the press an impression is made in

metal, but if the student is alert and observant, the most important impression is that which is made on his mind. The student must remember that no matter how adept he may become in forming sample pieces at school, this is not all that is required of him. The real purpose behind such training is to awaken his own reasoning powers, to develop his initiative so that he may be able to overcome more difficult problems when he tackles the real thing. It is the instructor's intention to form the young mind rather than the actual metal, and this is best accomplished when surrounded by the "atmosphere" of the workshop.

All instruction is free, funds for such training at present being supplied by the Federal Government as part of the national defense training program. In these schools, not only do the boys learn the most up-to-date industrial methods, but they are taught such general work habits as:

1. Proper attitude toward authority.
2. Dependable attendance.
3. Obedience in carrying out orders.
4. Initiative and resourcefulness in emergencies.
5. Neatness and accuracy of work.
6. Promptness in reporting for duty.
7. Reliability at all times and in all matters.
8. Desire to co-operate when working with others.
9. Constant attention to duty.

The foreman's first impression of a new employee is lasting. The apprentice with good vocational training is in a position to create a more favorable first impression.

C. Training in Industry—The present abnormal demand for aircraft has also forced manufacturers all over the country to attempt to train the young men who are already in their employ, but who lack classroom training. This phase of vocational training is also made possible through the use of Federal national defense funds. Young men who are employed on day-shift work attend school during evening hours, and those on night work attend during daylight hours. Such courses as aircraft blueprint reading, related shop theory, electrical blueprint reading, aircraft sheet metal layout and development, aircraft related English, shop mathematics, shop trigonometry, and industrial supervision are available to those apprentices who have the ambition to attend. In most cases the instructors are technicians from the aircraft factories, which minimizes the possibility of superfluous instruction being given.

An attempt has also been made by the aircraft industry to set up evening courses in tool engineering. This is particularly valuable for toolroom apprentices, as it provides them with the most logical first step in the direction of the tool engineering department. In these courses, tool and operation planning and tool-drafting procedure are studied. These subjects include the selection and heat treatment of steels, and a study of all the machines used in the aircraft industry.

The foregoing outline gives the prospective apprentice some idea of the valuable free instruction that is available in most of the aircraft centers, although in some locations the systems may vary slightly. In all cases, it is suggested that the boy should seek the advice of the guidance counselors who operate in various schools throughout the country.

D. Private Trade Schools—Another source of apprenticeship training is the private trade schools. Some of these are equipped to teach every single phase of the aircraft industry included in the design, manufacture and flying of the airplane, and also airport administration. In choosing this type of training, the boy has to consider the fees involved, as well as the integrity of the instruction given, so it is very important that he be advised by someone already in the industry before making his choice of a school. Such schools are located in almost all industrial areas.

For those who have definitely decided to follow aircraft

welding as a career, there are numerous trade schools that specialize in that field only.

In America today, the possibilities of gaining a sound pre-employment training are unparalleled, so you who are to be the builders of the aerial liners of tomorrow or the creators of our aerial navies of the future must take full advantage of all these facilities.

CHAPTER 3

GETTING A FOOTHOLD IN THE AIRCRAFT INDUSTRY

Getting a Job—After absorbing as much pre-employment vocational training as possible, the time approaches for the boy's entry into the aircraft factory. He feels excited because this is something new. When he entered high school he had no difficulty in enrolling, but now he must use his personality, his powers of persuasion. He must begin to introduce dollars and cents into his scheme of things. He must have a ready response to questions about his past training, and must be sure that his training qualifies him for the job he is after. He feels more of a man because he is out in the world to make his own way. "But listen, boy, you're not ready to dive off the deep end yet. You still need guidance and must look for a guiding hand. As far as you are concerned, the industrial sea is uncharted, and for some time you must be piloted through it."

The problem of the prospective apprentice now is to get his name on the roll of one of the many aircraft factories that dot the map from East to West. It must be understood, however, that vocational and high school training are not a guarantee of a job. Most of the aircraft companies have chosen a man to act as their Personnel Director, and it is for the apprentice to prove to this individual that the company will benefit by employing him.

The apprentice should indulge in a little self-analysis before coming into contact with the Personnel Director. A wrong approach or a bad presentation of his capabilities may spoil his chances. In thus applying a yardstick to himself, the apprentice must have the courage to keep his feet on the ground, otherwise his measure will not be accurate. His shortcomings will not be apparent.

Contact by Letter—Obviously, of course, there are two ways of getting in touch with the Personnel Director, namely, by letter or by personal contact. If properly executed they may be equally effective. The former method, however, should not be attempted unless the applicant is capable of writing a letter that will arouse the interest of the Personnel Director, or unless he is prevented by distance from making a personal visit. Such a letter should be written by hand (not typed), and the spelling must be correct. It should state clearly the type of work desired and must give a full account of the qualifications possessed by the writer relative to the job in question, also his educational background. It is often advantageous to mention hobbies. If he has any suitable tools, they should also be listed. A few names may be given as references, but they should be of professional men. Full scope should be given to accomplishment and training, but these must not be overestimated. The truth must be told. In many cases, well-written and well-arranged letters have brought offers of immediate employment, without the necessity of the applicant's visiting the factory. Nor is this method of contact denied to the boy who was born in a foreign land and whose knowledge of the English language may yet be in its initial stages. This is essentially a country of equal opportunity for all, and the average Personnel Director is one who is able to recognize intelligence, even if it is expressed in broken English.

Personal Contact—If immediate employment is not offered, an application blank will probably be forwarded. This should be filled in promptly and in such a manner that the type of work desired is specifically noted, and that any vocational training is emphasized. If possible the application form should be delivered personally rather than by mail. This affords the applicant a valuable opportunity of using his personality where his previous letter failed.

The boy who realizes his shortcomings with the pen should curtail his letter writing to a mere request for an application blank. When he calls at the employment office in person with the filled-in application form, this boy's chances of being hired depend entirely upon how he impresses his interviewer. The things he should stress are: willingness to work hard, enthusiasm for the particular job he has in view, his past training in that line, his capacity for learning more, and his willingness, if necessary, to accept another job until a vacancy occurs in the department of his choice.

From the moment his name is entered on the company pay roll, the boy from vocational or high school must realize that his status in life has changed. He has become a vital part of a great industry, and as such he must give as well as receive. In fact, his success will be assured only if he gives not as little but as much as possible.

Apprenticeship Systems—The apprenticeship system, whereby an agreement is drawn up and signed by the company on the one hand and the apprentice and his parent or guardian on the other, is not in use to any great extent in the aircraft industry. In some factories it is used, involving only a limited number of boys who are chosen from those already employed. The general procedure is to employ the boys as helpers at a low starting salary, to put them to work with skilled men, and to raise their salaries as their usefulness increases. Nor is the boy duty bound to stay in one department, as transfers from one department to another may be arranged upon request. Very seldom will the right-thinking foreman stand in the way of an apprentice who wishes to expand his knowledge.

The boy who thus starts out with the idea of gaining a foothold in the aircraft industry should seek training along two separate lines. In the first place he must train for his practical work (which, of course, includes the necessary theory), and secondly he should train himself in the art of leadership. In this way he can prepare himself for promotion. The foreman he works for knows just what qualities a leader should possess and how to recognize them, so it is to the advantage of the apprentice to develop such qualities.

1. Physical—The good foreman is somewhat of a psychologist, so he analyzes his men from three points of view, namely, the physical, the mental and the moral. Naturally, if he is to be a successful leader the boy must take care of his health. He must be capable of day-in-day-out service without much interruption. Also in some cases, a sickly body undermines the state of mind. This creates an inferiority complex, which destroys all the other qualities of leadership.

2. Mental—The mental capacity of the apprentice is also under close observation. The foreman wants to find out his attitude towards his job. Is he enthusiastic? Does he wait for the quitting bell as a runner would the report of the starting pistol? Does he have a quick perception and an open mind? Is he anxious to learn more? Does he understand company policies and adhere to company rules? These are the things that the apprentice should think about, so that the foreman may recognize in him the qualities of leadership.

3. Moral—That the prospective leader should practice moral integrity is exceedingly important. It is easy to lead others from a higher moral level; they can see and respect the example so displayed. On the same level as his subordinates, the leader loses much of his power; he descends to a more obscure position. From below that moral level, however, leadership is utterly impossible.

Good Language—In order still further to enhance his position, the apprentice should learn to become expert in his use of the English language. When he becomes a leader it will be necessary for him to write reports and formulate plans which he must submit to his superiors. Written matter of this nature must be clear and concise, leaving no room for misapprehension. He must also learn how to talk so that his words will carry weight. This does not mean being talkative. To the contrary it is better to use as few words as possible, but to choose the correct ones. The apprentice should strive to increase his vocabulary, so that he may have a ready choice of words. His reading matter should be chosen wisely, and a large portion of it should be of a technical nature. In this way he will keep abreast of the times. To help him in his choice of books, the following list has been compiled. Many other valuable books, however, may be had from any of the various libraries throughout the country.

Almen, H. V., and Mead, R. K., *Aircraft Blueprint Reading.* Pitman Publishing Corp., New York. Revised, 1942.

Ashkouti, J. A., *Aircraft Mechanic's Pocket Manual.* Pitman Publishing Corp., New York. Revised and enlarged, 1942.

Barber, Capt. H., *The Airplane Speaks.* R. M. McBride & Co., New York.

Bartholomew, N. C., *Aircraft Inspection Methods.* Pitman Publishing Corp., New York, 1940.

Boggess, H. E., *Aircraft Sheet Metal Work.* Pitman Publishing Corp., New York, 1941.

Brimm, D. J., and Boggess, H. E., *Aircraft Engine Maintenance.* Pitman Publishing Corp., New York, 1939.

Brimm, D. J., and Boggess, H. E., *Aircraft Maintenance.* Pitman Publishing Corp., New York, 1940.

Carter, C. C., *Simple Aerodynamics and the Airplane.* Ronald Press, New York.

Ciffrin, A., and Von Borchers, C., *Aircraft Torch Welding.* Pitman Publishing Corp., New York. Revised, 1942.

Downer, A. E., *Practical Mathematics of Aviation.* Pitman Publishing Corp., New York. Third Edition, 1942.

Hume, D. C. M., *Elementary Aerodynamics.* Pitman Publishing Corp., New York, 1941.

Kutakoff, B. A., *Plane Facts for Airplane and Engine Mechanics.* Military Book Co., 1941.

Markey, R., *The Aircraft Propeller.* Pitman Publishing Corp., New York. Second Edition, 1942.

Marschner, C. F., *Aircraft Assembly.* Pitman Publishing Corp., New York, 1942.

Nolan, R. R., *Machine Tools in Aircraft Production.* Pitman Publishing Corp., New York, 1941.

Norcross, Carl, *The Aviation Mechanic.* Whittlesey House, New York, 1941.

Nye, W. L., *Metal Aircraft Design and Construction.* Aviation Press, 1938.

Thompson, James G., *Aircraft Drafting Room Manual.* Aviation Press, 1939.

Titterton, G. F., *Aircraft Materials and Processes.* Pitman Publishing Corp., New York, 1941.

Duties of an Apprentice—In most of the large corporations, the new employee receives a book setting forth the company policies and rules with which the apprentice should familiarize himself. In this way he realizes that certain duties and obligations are expected of him. For instance, all his time must be properly charged and recorded; smoking rules must be observed or infringements penalized; punctuality is expected and company property must be respected. By knowing the company rules, the apprentice can better keep out of trouble.

Interest in Job—There are certain duties and obligations that the apprentice owes to himself if he has foresight enough. He must not lose sight of the fact that he voluntarily asked for employment so that he might learn to become more proficient in the job he chose as his life's work. This necessitates his strict attention to his job and to the instruction he receives from his co-workers who are more highly skilled than he. It has been observed that some apprentices go to work with the intention of taking it easy. They think it smart to dodge the foreman and to work only when they are being watched. This only hastens their downfall, however, because the average foreman knows all the tricks of the apprentices. He also knows human nature, so he usually warns the erring apprentices, thereby giving them another chance. Those who think that the foreman can still be trifled with usually find themselves without a job and plenty of time to take it easy with no foreman to dodge.

The apprentice must also guard against becoming bored with his job. The tendency is for the air-conscious youth, who is enthusiastic only about the romance surrounding the completed airplane, to minimize the importance of the little job to which he may be assigned. He becomes, to his way of thinking, hopelessly enveloped in an army of other workers. The everyday job of drilling and riveting, the rat-tat-tat of a wooden mallet on a piece of aluminum hold no attraction for the boy just out of high school—especially when he remembers that somewhere at the other end of the factory the extremely fortunate ones are wheeling the completed airplanes out for test.

They are the ones, he grumbles, who really learn something about airplanes. It is they who feel the take-off slip stream in their hair, who talk with the pilots about ceiling and visibility or "giving her the gun." Listen, boy, forget all this. You're going to send that test pilot to his doom if you don't attend to your job and quit dreaming. That's how important you are. The placing of every little rivet you insert has been studied by a stress mathematician. It has been provided for by a competent tool designer and located for you by one of those remarkably accurate tool makers, who split an inch into a thousand pieces and know what to do with every piece. So it's up to the apprentice to finish the job by driving home a sound rivet, one that will hold when there is a long flight ahead and the bulkheads are being stressed to the limit under a heavy gasoline load. In this industry there is a constant search for young men who possess integrity, initiative, and, above all, patience and perseverance at detail work, so that they may be given more and more responsibility.

Due to the fact that the airplane must possess at the same time both strength and lightness, every detail in the structure is designed and placed to withstand its own definite load only. There is not much room for overlapping, hence every detail must be fabricated strictly according to specification, otherwise its failure under stress may well crash the plane and prove fatal to human life.

"There's your responsibility young man. Do you accept it as your contribution to the aircraft industry?"

To acquaint the prospective apprentice with the different types of jobs that are available in the aircraft factory, the occupational chart in Fig. 1 has been compiled.

Before starting to work in any of the manufacturing departments, the apprentice should obtain certain necessary tools, and he will do well to get the best available. If he is so fortunate as to have an experienced mechanic as a friend, he should ask advice. Cheap tools are no bargain.

The necessary starting tools will be listed as each department is discussed.

Safe Practices—In the interest of safety, the apprentice must be always on the alert. When applying rope slings to any object for lifting with a crane, care must be taken to see that the center of gravity of the object is low enough to prevent its toppling when lifted off the ground. The type of knot used is also very important, in order to prevent slippage. In this respect the apprentice should find out from his foreman how to make the simple reef knot. When trucks have been specially designed to carry specific loads, they should not be used to carry any other type of load. The zealous boy should also refrain from lifting by hand objects which are heavy enough to cause strain. It is safer to ask one's neighbor for help. A flowing necktie near a whirling spindle is a very grave source of danger. When caught, the necktie rolls around the spindle so fast that the result is sometimes very gruesome. Sleeves should be rolled up, as they constitute a similar danger source. Very often, the drill press operator, intent on his job with head bent

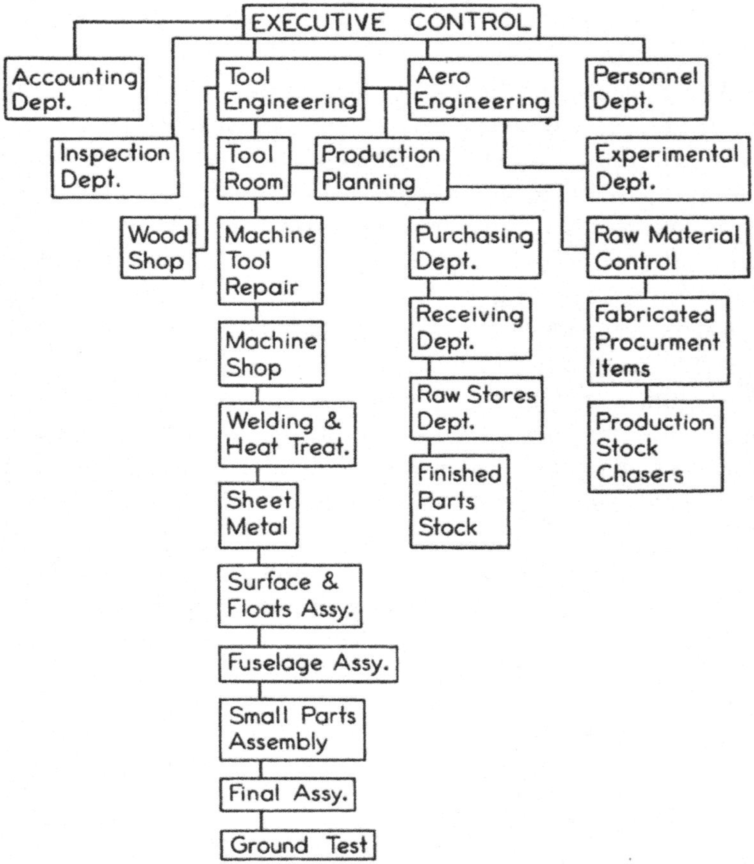

Fig. 1—Occupational Chart

down, is shocked to find that his hair has been attracted towards the revolving spindle, subjecting him to a sudden loss of his forelocks. Sometimes part of the scalp goes with the hair, resulting in a very painful type of accident.

If tonsorial fashion decrees long hair for boys, it is better for the apprentice to be out of fashion if he works on a drill press.

Another type of drill-press accident is caused by the job's not being held securely enough. The drill catches, the job flies around, and woe betide any fingers that may be in its path.

When using the bench or floor grinder, the eyes should be protected with goggles from flying particles of steel or carborundum. When the hand grinder is used, a screen should be erected to protect others in the vicinity. Also in the case of the hand grinder, care must be taken that the wheel diameter is not excessive. The centrifugal force which is set up by these tools is so great that, if the proper speed-diameter relationship is not maintained, the wheel may fly apart, causing serious injury.

At lunch or quitting time, it is dangerous to run. A spot of oil on the floor may precipitate an undignified and injurious somersault. Short cuts through the factory are dangerous, so aisles have been designated and must be used.

If the apprentice respects his chisel hand he must learn to look at the working end of the chisel and not at the head.

Any form of practical joke is unconditionally prohibited within the factory. Cold steel has no sense of humor.

CHAPTER 4

ELEMENTS OF THE THEORY OF FLIGHT

In order to help the apprentice further in gaining a foothold in the aircraft industry, there follows a description of what is happening around him, and how he himself fits into the picture. As his attention to his own duties prevents his acquisition of this knowledge, this and the following chapters are dedicated to that end. Not that the man in the shop is expected to be an engineer, draftsman and mechanic rolled into one, but a knowledge of these things will secure his interest more firmly, increase his value to the company, and ensure his more rapid promotion to greater responsibility.

Aeronautical Engineering Department—It has always been rather difficult for the apprentice in the shop to realize the importance of the aeronautical engineering department. This is due to youth's tremendous capacity of taking things for granted. He receives and studies the blueprint of the part he is expected to produce, but as to its origin or the procedure that brought it to him he has little curiosity. Had his foreman told him that it was designed, drawn and printed by a machine, he would probably think that in an aircraft factory anything is likely to happen.

The ultimate goal of the engineer, after each piece has been drawn up and its stress resistance calculated, is to produce an airplane that has good flying characteristics. In order to stimulate in the youth an interest in this field, let us study the basic rules of flight.

To the average person, the fact that a machine weighing many tons can fly through the air by its own power is nothing short of miraculous. Such an open defiance of the law of gravity, some say, is an encroachment on the supernatural and is bound to lead to disaster.

Lifting the Airplane—To the contrary, however, the transformation of the air of the atmosphere into a very powerful supporting medium is the result of some very natural conditions. This does not necessitate a change in the substance or chemical value of the air but is due entirely to a difference between the relative motion of the airplane and the air itself. The following simple experiment will illustrate this point very clearly.

The next time you ride as a passenger in an automobile, put your hand out into the air stream. Incline your hand upward, so that the air meets your palm, and you will find that your hand is forced upward. This upward pressure increases with the speed of the car. Were you to hold a two-foot by four-foot board out in the same air stream and incline the front of it upward, the pressure would be greater than your capacity to hold the board in place. This proves that the upward pressure also increases with the area of the inclined plane.

FIG. 2

Angle of Incidence—In the earlier types of aircraft, the wings were little more than boards jutting out into the air stream. The upward tilt which gathered the pressure under the wing was called, as it is today, the "angle of incidence" (Fig. 2). The early pioneers also discovered that by depending entirely upon the angle of incidence for their lifting power, the air would continually pile up below the wing, and thus destroy the even flow necessary for maintaining an even, upward pressure.

Modern Airfoil—This was overcome by the use of a wing which produced lift more by the flow of air over it than by the contact of air below it. This is the airfoil which is used in varying forms in the aircraft of today.

When a wing of this type is in operation the air is thrown upward, causing a low-pressure area (Fig. 3) above the wing. Since nature abhors a vacuum, there is a tendency for the surrounding elements to close in and equalize the difference in pressure. The wing itself, therefore, is subjected to a constant upward pressure during flight in order to fill in the low-pressure area above it. The vertical line through the airfoil at which this pressure is greatest is called the "center of lift."

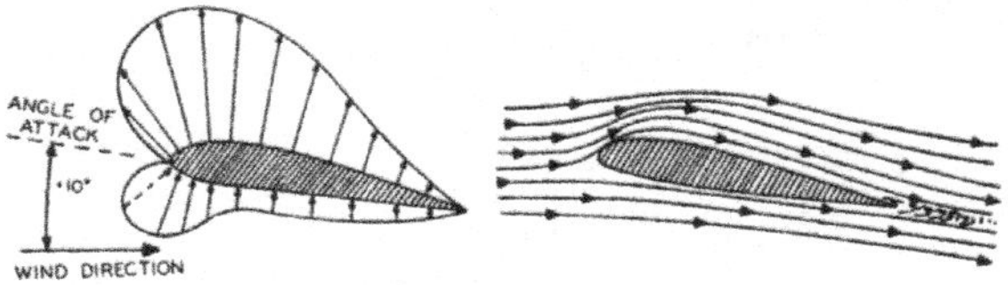

Fig. 3—Pressure and Flow at Medium Angle of Attack

Any interference with the action just described will seriously affect the flying characteristics of the "ship." For instance, if the airplane is flying in very low temperatures, ice may form on the leading curve of the wing. This will destroy its shape and reduce the upward thrust of air to such an extent that the low-pressure area will vanish, causing the airplane to fall.

Weight—As can readily be imagined, the engineer must be very careful as to how he distributes the weight of his airplane. Extreme tail heaviness or nose heaviness must be avoided at all costs. He must know the exact weight of every detail and on his complete assembly drawing must place everything in such a way that, when in flying position, the center of lift is just behind the center of gravity. With this condition prevailing, should the engine be shut off in flight, the airplane will go into a natural glide, with all the surfaces meeting the air in the manner in which they were designed. This permits effective control in a glide to *terra firma* without the pull of the motor. When an airplane is thus able to maintain automatically a normal attitude, it is said to possess "inherent stability."

Dihedral—Another feature which tends to produce such stability is the "dihedral angle." This is seen when looking at the airplane head on (as in Fig. 4), and is the upward

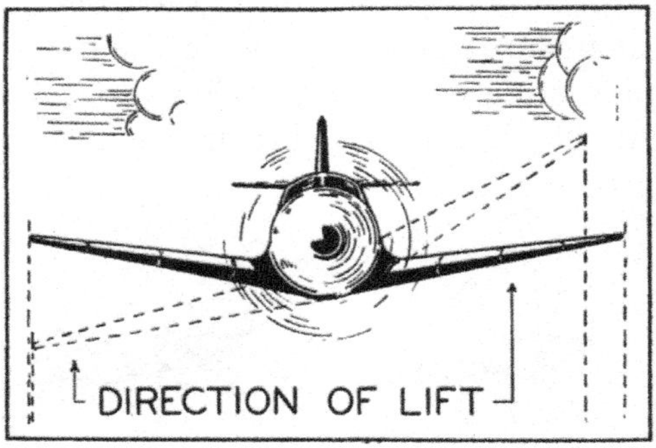

FIG. 4

angle of the wing from its root to its tip. Suppose the airplane assumes the position shown in dotted lines; observe that the length of the lifting surface is longer on one side than the other. The short side, thus receiving less lift than the long side, automatically drops until a balance is reached.

Control—Now that we are able to lift the airplane, the question of controlling its movements in the air arises. Control, in this case, may be termed: the ability of the aircraft to change direction of motion or speed at the will of the pilot, while still maintaining stability.

In comparison with all other forms of transportation, the airplane is subject to far more different forms of control. Consider, for instance, the ability of the airplane to climb, dive, bank, turn, loop the loop, or even fly upside down. In order to accomplish these maneuvers, three different systems of control are necessary, and it is essential that they be applied in close co-operation with one another.

1. Throttle Control—First of all is the throttle, which regulates the revolutions of the engine; secondly is "lateral control" (Fig. 5); and thirdly is "longitudinal control"

FIG. 5

(Fig. 6). As the next chapter is devoted to the internal-combustion engine, we shall consider at this time only the systems of lateral control and longitudinal control.

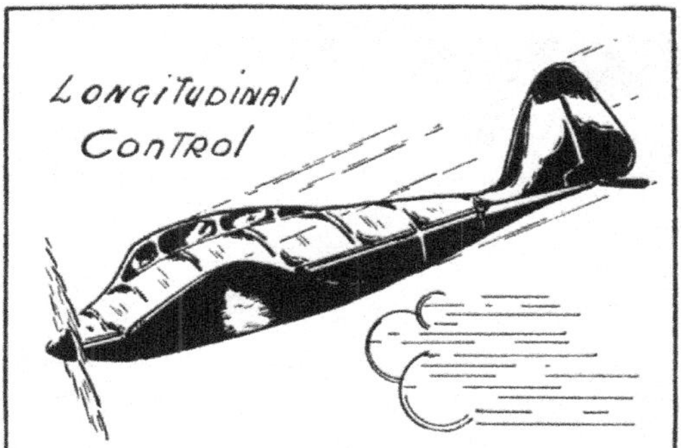

FIG. 6

2. Lateral Control (Ailerons)—As shown by the sketch (Fig. 5), lateral control is the ability of the aircraft to change its position from wing tip to wing tip, in relation to the horizon towards which it is flying. This is accomplished by means of the "ailerons," or hinged portions of the trailing edges of the wings. In operation, when the right aileron is hinged upward, the left one is in the down position. This causes the left wing to be forced upward because the lowered aileron obstructs the passage of air beneath the wing, creating a pressure. The right wing, on the other hand, is forced downward, owing to the reversed position of the aileron, giving the airplane a lateral movement. This movement is obviously converted into a banked turn by the use of the rudder.

3. Longitudinal Control (Elevators, Stabilizer)—Longitudinal control is the ability of the airplane to change its position fore and aft, and is accomplished by means of the elevators. These are located in the tail of the airplane and are hinged to the stabilizer (Fig. 7).

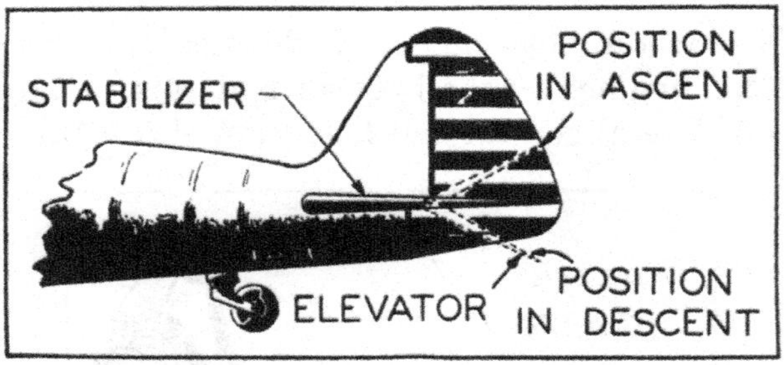

FIG. 7

The normal flying position in flight is maintained by the stabilizer, as the tail will move automatically up or down until the pressure above and below the stabilizer is equalized. If, however, the elevator is moved, the equality of pressure is immediately destroyed, and the tail has to alter its position to restore the equilibrium. The airplane has then altered its position longitudinally. The ailerons and the elevators are controlled by the movements of the "stick control." This is the main control lever of the airplane. In the neutral position it stands vertically upright from the floor of the pilot's cockpit, high enough for the pilot to grasp comfortably. It is hinged at the bottom, so that it can be swung through any arc. A movement of the control stick in any particular direction will produce a movement of the airplane in the same direction. For instance, if this ingenious device is pulled backward towards the pilot, the airplane will do just what would be expected from such a movement—it will go into a climb. A forward movement away from the pilot will produce a dive. A move to the left, supplemented by the rudder, will cause the ship to bank in that direction, and vice versa.

Varying combination maneuvers can be produced by the in-between movements of the stick. Another type of stick control utilizes a handwheel for moving the ailerons, the stick itself being restricted to a backward and forward movement in order to control the elevators.

Construction—Besides providing for the mechanical perfection of the moving parts of the airplane, the engineering department is charged with the responsibility of designing a structure that combines strength and lightness. This is no easy job, and it has taxed the ingenuity of designers the world over. In the early days of the industry most of the stress members of the airplane were made of wood, being braced with wires. Soon after the Great War of 1914-18, however, it became almost the standard practice to use the welded steel tube construction for fuselages. This, in turn, has given way to the almost universal use of aluminum and its alloys, a combination of metals possessing phenomenal strength for their weight. These metals are available as castings, sheet and bar stock, and in extruded forms. The latter is proving of immense value to the builders of aircraft. The extrusion, as it is called, is made by forcing a billet of metal through an aperture in a hardened steel die. The metal that emerges takes exactly the same shape as the hole through which it has been forced. In this way, long bars of assorted shapes are obtained, which require no machining except cutting to length and drilling, if necessary.

The various shapes are decided upon by the stress analysts of the engineering department, each radius, thickness of web, or height of flange being closely figured to add maximum strength coupled with minimum weight to the extrusion. So you see that in the design of fuselage and wing, the engineers must maintain an outward form that will ensure efficient flight, at the same time creating an internal structure of bulkhead and rib that will be more than able to withstand the enormous stresses that are set up by modern flying practices.

Stresses—Think of the built-in strength of the modern dive bomber that comes roaring out of the sky at a speed of 450 miles per hour. Almost vertically downward he carries a ton of bombs which he releases at a height of a few hundred feet, and then immediately zooms upward. Have you ever felt the stresses set up in your automobile by a sudden change in direction? Have you ever seen a passenger dive through a windshield due to a sudden loss of forward motion? Conservative or reckless driver, you must surely realize by these comparisons the tremendous stresses that are set up in modern flying.

CHAPTER 5

THE PRINCIPLES OF THE AIRCRAFT ENGINE

Flight, as described in the last chapter, requires power in order to produce forward motion, and this is supplied by the engine. The propeller, or airscrew, being rotated at a high rate of speed by the engine, "screws" itself into the atmosphere, causing the whole airplane to move forward.

There are two distinct types of engines that are suitable for aircraft: one is the Diesel oil engine and the other the gasoline engine. Both of these engines are of the internal-combustion type, due to the fact that the fuel is burned or ignited within the cylinders of the engine itself. This is in contrast to the steam engine in which fuel is burned at the boiler. The immense quantities of water as well as fuel make the steam engine unsuitable for aircraft use.

DIESEL ENGINES

Although the Diesel engine is not the type that is generally used in aircraft, it has some decided advantages over the gasoline type. In the first place, it has no complicated electrical ignition system, as in the gasoline engine. The ignition of the fuel is attained by merely injecting a stream of oil at high pressure into the compressed air in the cylinder head. The combustion is spontaneous. The oil immediately vaporizes and expands, due to the heat of compression, driving the piston down on its stroke.

GASOLINE ENGINES

As the gasoline engine has a more favorable power-weight ratio than the Diesel and has been developed to a point where it is very dependable, it is at present the accepted source of power for aircraft.

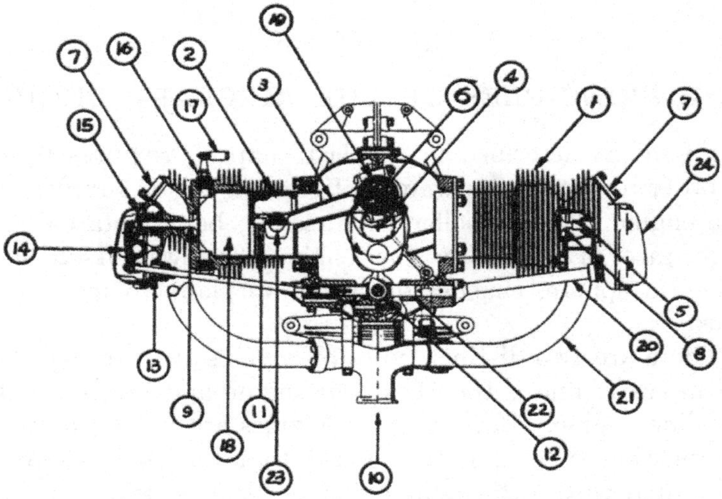

From "Aircraft Engine Maintenance" by Brimm and Boggess

FIG. 8—HORIZONTALLY OPPOSED CONTINENTAL A-50

1. Cylinder
2. Piston
3. Connecting rod
4. Crankshaft center
5. Intake port
6. Crankshaft rotation
7. Exhaust port
8. Intake valve
9. Exhaust valve
10. Carburetor intake
11. Piston rings
12. Camshaft
13. Pushrod
14. Rocker arm
15. Valve spring
16. Spark plug
17. Spark plug terminal
18. Combustion chamber
19. Crank pin
20. Pushrod housing
21. Intake pipe
22. Cam follower
23. Piston pin
24. Rocker box cover

Pistons and Cylinders (Power)—It is important at this point that the apprentice should understand fully the principle of the piston and the cylinder. Supposing we take a metal cylinder, closed at the bottom, whose inside wall is perfectly round and smooth. Into this cylinder is poured some water, and a piston or round metal disc is inserted so that it rests on top of the water. The piston, of course, must be a neat sliding fit in the cylinder. If heat is then placed under the cylinder, the water will reach boiling point. Steam is thus generated and there being no escape, the pressure under the piston begins to mount. This pressure, or power, will eventually overcome the weight of the piston and blow it right out of the cylinder. The foregoing is an example of the generation of power. In the engine —whether steam or internal combustion—the piston does not fly off into space, but it comes back for more, and the power is collected and delivered to the propeller or wheel, as the case may be. In the gasoline engine, a connecting rod is attached to the piston at one end, and to a crank pin at the other. This naturally restricts the movement of the piston, giving to it a reciprocating motion which makes possible the very definite strokes which we shall now discuss.

In order to co-ordinate the movements of the piston with the entry of the fuel into the cylinder, its ignition and final exhaust, the four-stroke principle has been devised. Sometimes this is called the "Otto cycle of operations," being named after Otto, the German inventor.

Four-stroke Principle—Let us study these operations as shown in Fig. 9. In view A, the piston is descending in the cylinder, the inlet valve is open and the gasoline vapor is rushing in to displace the piston. The latter soon reaches the bottom of its stroke, and in view B it is ascending. The inlet valve is closed, the mixture is trapped and is compressed. At the top of the stroke the ignition occurs, the heat of which causes the mixture to expand very rapidly, driving the piston down on the power stroke, as in view C. At the end of this stroke the exhaust valve opens and the piston returns, driving the burnt gas out through the open valve (view D). After this the inlet valve opens, the exhaust closes and the cycle of operations is ready to repeat. Thus, it will be seen that it takes two revolutions of the crankshaft to complete the four strokes: inlet, compression, power, and exhaust.

In an engine with only one cylinder, the momentum of the flywheel keeps it running during the three idle strokes. With the addition of another cylinder, a power stroke is obtained for each revolution, or two power strokes during the cycle of operations. This increases the efficiency of the engine twofold. In the case of the multi-cylinder engines that are installed in the aircraft of today, the number of cylinders equals the number of impulses transmitted to the crankshaft during two revolutions. This is one of the reasons for the tremendous power that it is possible to generate in these engines.

A B C D

From "Aircraft Engine Maintenance" by Brimm and Boggess

FIG. 9

A. Intake stroke (intake valve open, exhaust valve closed)
B. Compression stroke (intake valve closed, exhaust valve closed)
C. Power stroke (intake valve closed, exhaust valve closed)
D. Exhaust stroke (intake valve closed, exhaust valve open)

Crank Arrangement—It is essential, however, that the crank throws of a multi-cylinder engine be set at the proper relative angles to each other, so that when each power stroke occurs, each piston will be in the proper position to receive the impulse. Fig. 10 illustrates the crank arrangement for a four-cylinder engine.

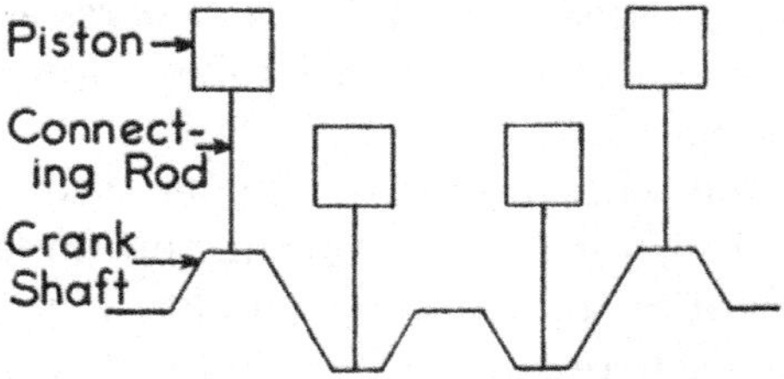

FIG. 10

Order of Firing—The order in which the power strokes occur, or as it is generally called, "the order of firing," is also very important in connection with the crank arrangement. A study of the four-cylinder engine in Fig. 10 will demonstrate this, if you remember that four power strokes occur during the two revolutions of the cycle of operations. If the cylinders are numbered from one to four consecutively, what is their order of firing? Suppose that number one piston is going down on the power stroke; this causes numbers two and three to ascend to the firing position. A check-up on the valves discloses which one of these two is the next to fire. The cylinder with both valves closed is the one to fire, say in this case it is number three. This brings one and four into position, but number one already has fired, so it must be number four. Verify this by checking the valves. The only cylinder left to fire now is number two, which duly comes up to the firing position with both valves closed.

From this we see that the order of firing of a four-cylinder in-line engine may be either 1,3,4,2, or 1,2,4,3, according to the valve openings. In a six-cylinder engine, the order of firing is usually 1,5,3,6,2,4. If the cylinders were to fire consecutively, the vibration would be excessive.

Valves—The control of the valves is an exceedingly important matter in the operation of the internal-combustion engine, and their openings must be closely timed in conjunction with the piston movement. The camshaft is principally responsible for the behavior of the valves, and it consists of a shaft which carries a cam or eccentric for each valve. The revolving speed of the camshaft is one half that of the crankshaft, because each valve opens once only for every two revolutions of the crankshaft. In order to time the valves, turn the engine until number one piston is in position for the inlet valve to open, then turn the camshaft independently in the direction of normal rotation until the inlet valve for that cylinder starts to open, then mesh the gears together. All the other valves are then automatically timed, because the camshaft has been made in one unit, each cam being set in the correct relationship to the other.

The valves themselves are of the mushroom type, with conical seats as shown in Fig. 11. Each valve has to be ground to fit its own seat, so that it will be absolutely free from leaks during the period at which the pressure is at its highest, namely, during the power stroke. The eccentricity of each cam as it rotates causes a rod known as the tappet to reciprocate, which in turn opperates the valve. So as to ensure that each valve will close completely, the tappets are adjustable. Obviously, any lengthening of the tappet will tend to open the valve and vice versa.

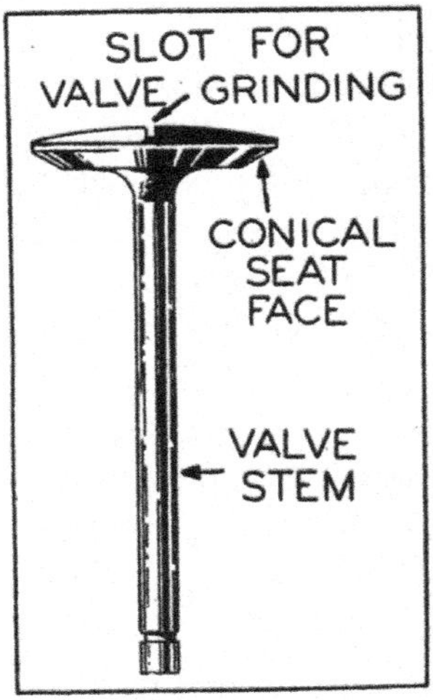

FIG. 11

Valve Clearances—In operation, the tappets must be set with a certain amount of clearance, usually specified by the manufacturers. It is the general practice to give the exhaust tappets more clearance than the inlet, so as to compensate for the extra expansion of the exhaust-valve stems due to the heat. So important are the tappet clearances, they are specified by the manufacturers in thousandths of an inch and must be set accurately with a feeler gauge.

Too much clearance will cause the valves to open late and close early, thus shortening the effective inlet and exhaust strokes. As the piston speed is much greater than that of the valve itself, a late opening or an early closure will occur with the piston well out of the specified position.

Cycle of Operations—Fig. 12 is a diagrammatic view of the complete cycle of operations for an average, medium-speed engine in which the curved line is traced through the two revolutions of the crank pin, while the valve openings and the ignition point are placed in their respective positions.

It will be seen that the inlet valve opens about 15 deg. past top dead center and 10 deg. after the exhaust closes. This 10 deg. of piston movement with both valves closed creates a slight subatmospheric pressure in the cylinder so that when the inlet valve opens, the air is induced through the carburetor and into the cylinder with an extra rush. This helps carburetion, which will be described later.

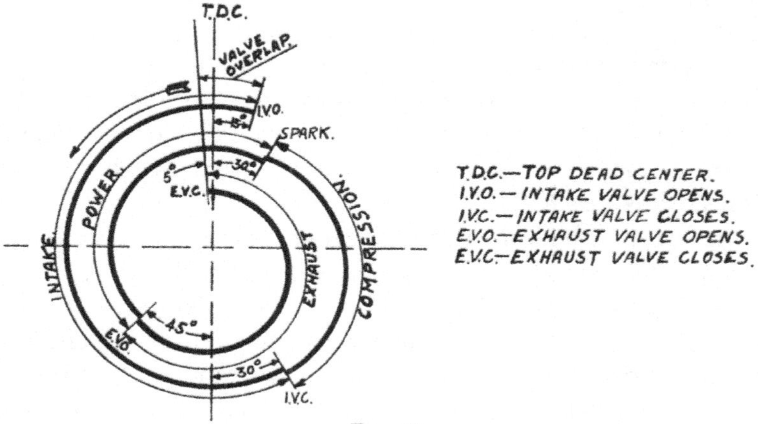

Fig. 12

The stroke then proceeds and the inlet valve closes at a point 30 deg. past bottom dead center. Instead of the piston's pushing some of the charge back through the inlet valve, as would be expected during this 30 deg. of the upstroke, the incoming gases are allowed to catch up with the piston, which has descended faster than the gases themselves. In this way the maximum charge is induced, for although the piston is ascending, the charge is still entering through the open valve.

With the closing of the inlet valve, the compression stroke begins. At a point approximately 30 deg. before top dead center, the spark takes place, igniting the mixture of gasoline vapor and air which has been drawn in and compressed. This advance of the spark is very important, as it allows time for the mixture to become fully ignited. When the piston reaches the top of the stroke, expansion is ready to exert its full pressure.

Were the spark to occur on top dead center, the full pressure would be attained with the piston already on its way down on the power stroke. As the piston speed increases greatly toward the middle of the stroke, much of the expansion would thus be wasted in just catching up with the piston.

At a point 45 deg. before bottom dead center, the exhaust valve opens. Although this sacrifices some of the pressure that would be applied to the piston, this disadvantage is more than compensated for by allowing the partly burned gases to escape under their own pressure. Much of the heat is thus dissipated.

During the remainder of the stroke, the cylinder is scavenged. The exhaust valve remains open until 5 deg. past top dead center, allowing time for complete expulsion of the products of combustion. The gasoline or fuel for the engine is stored in the tank in the form of a liquid, but it must enter the cylinder as a vapor. For this reason it must pass through the carburetor, Fig. 13.

Carburetor—As the presence of oxygen is necessary in producing combustion of any kind, an air intake is provided through which, at every inlet stroke, is drawn a charge of air. In the middle of this air passage is placed a jet with a very small hole, through which flows a thin spray of liquid gasoline. On its way to the cylinder through the inlet pipe, the gasoline vaporizes and unites with the air to become a highly explosive mixture. It must, however, pass through the throttle valve, which is merely a partition that can open or close the inlet passage at the will of the pilot.

With the engine at rest, the gasoline is maintained at a level slightly below the tip of the jet. The start of the engine creates a suction at the jet which draws the gasoline out in a spray. In order to increase the velocity of the air at this point, thereby atomizing the gasoline to a higher degree, the inside of the air passage is shaped as a "venturi tube." By restricting the air passage gradually, a higher pressure is attained; then by suddenly increasing the diameter of the passage, a high velocity of air results.

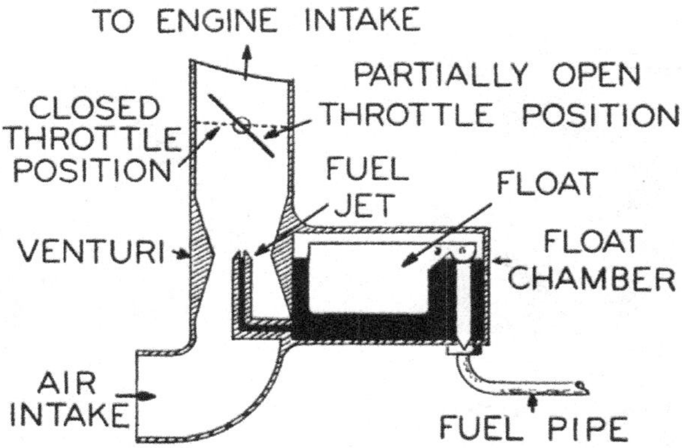

FIG. 13

The level of gasoline at the jet is maintained by means of the float chamber. This ingenious but simple device is fed with gasoline by the pump. Upon reaching a certain level, a float which rises with the gasoline closes a needle valve, shutting off the supply. The jet is connected with the bottom of the float chamber. As the tip of the jet is slightly above the level in the float chamber, there is always a column of gasoline in the jet, corresponding to this level and ready to be drawn out as the engine turns.

Ignition System—The ignition system in the aero engine is of the magneto type, and a brief description of this is essential. The magneto is merely a generator of high-voltage alternating current, which delivers its voltage to each spark plug at precisely the required instant.

The electromotive force is generated in an armature, on

the principle that if a copper wire is rotated in a magnetic field, as in Fig. 14, a voltage is produced in the wire. The voltage is proportionate to the number of lines of magnetic forces cut by the wire, and the speed with which they are cut. This means that the voltage increases with the strength of the magnets and the speed of the armature, the latter being a soft iron core around which is wound hundreds of coils of wire and which rotates in the field of the magneto magnets.

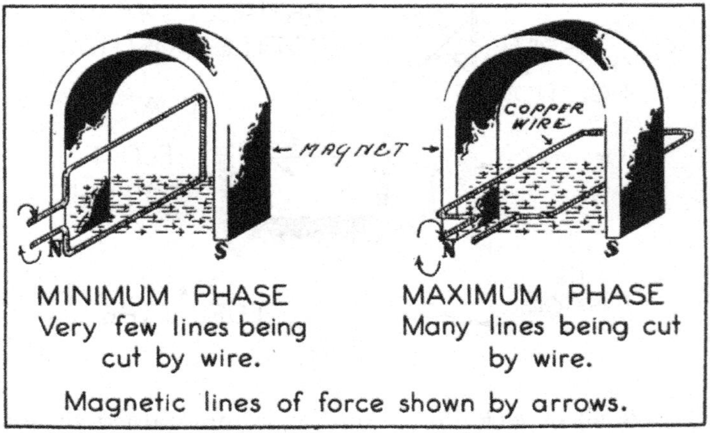

FIG. 14

The voltage thus generated is very spasmodic, due to the fact that, during two phases of each revolution, a vast number of lines of forces are being cut, while during the other two phases, very few are being cut. At its height the voltage reaches 30,000 and at this point it is conducted to the spark plug.

On the armature are two different windings, namely, the primary and the secondary. The former is composed of a few turns of thick copper wire—the latter, many turns of very thin wire. As the armature revolves, a low voltage is generated in the primary winding—one end of which is grounded to the engine, the other end being connected to one of two interrupter points. These points are really two small, round discs of platinum, one of which is movable, being actuated by a cam. This movement causes the points to open and close. When closed, the primary winding is completely grounded to the engine. One end of the secondary winding is connected to the same point as the primary, the other end being connected to the distributor terminal. The resulting action is one of intense electrical activity. The armature turns, a low voltage is generated in the primary, the points break, the low voltage of the primary surges through the countless coils of the secondary, which steps up the voltage to approximately 30,000.

The current is then conveyed to the distributor terminal of the magneto, so named because it distributes the current to each cylinder. This is done by means of a rotating carbon brush which sweeps past and touches one by one a number of brass segments. Each segment is connected by a wire to a spark plug, the number of segments, therefore, corresponding with the number of cylinders in the engine.

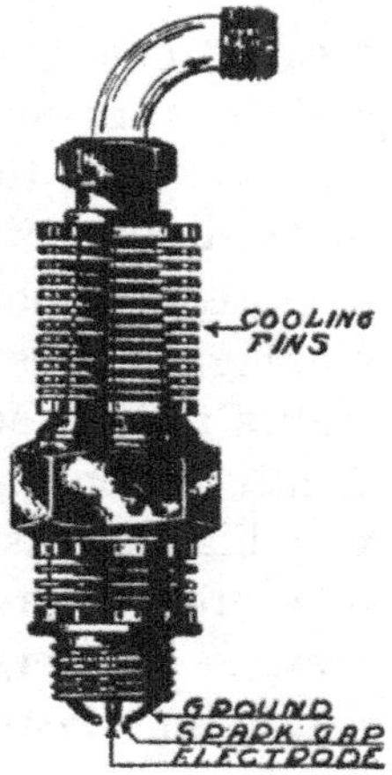

FIG. 15

Spark Plugs—At the moment of contact between the carbon brush and the segment, the current is immediately conveyed to the central electrode of the spark plug (Fig. 15). This electrode is heavily insulated from the threads of the spark plug, and therefore from the cylinder itself. It extends, however, down into the combustion chamber. The electrical current at this point is confronted by the spark gap, which it jumps, causing a spark to occur. The fuel is thus ignited. When the current jumps the spark gap, it completes the electrical circuit with the end of the primary winding, as both are grounded to the engine.

In starting an internal-combustion engine, it is necessary

to make it revolve from an outside source of power, in order to draw in fuel and ignite. The engine will then keep running, the pilot opening or closing the throttle valve to produce the desired number of revolutions. Engines of low horsepower can be started by swinging the propeller by hand, but larger engines must utilize an electric starter powered from the battery.

To stop the engine, the pilot closes the short-circuiting switch so that the primary current cannot be broken. This prevents the high magnetic flux through the secondary, thus making it impossible to spark at the plug.

Lubrication System—The apprentice who wishes to engage himself with things mechanical must realize the importance of proper lubrication. Where metal comes in moving contact with metal, friction is the result, and friction is the greatest enemy of the smooth-running machine. It wears and tears the moving surfaces until they actually adhere to one another; then motion ceases altogether. Oil is the magical fluid which defies friction. Being composed of microscopic globules, the oil spreads out thinly between the moving surfaces with a smooth-rolling action that actually makes the surfaces improve with use. The aero engine is vitally dependent on proper lubrication, as it operates at very high speeds and is subjected to very high working temperatures. The oil is stored in a tank and is pumped under pressure to every working part.

Cooling System—In every internal-combustion engine, the heat of combustion must be dissipated, and this presents a problem to engineers. In some engines, usually those whose cylinders are in line or in a V arrangement, the cooling is done by means of a liquid. Each cylinder is surrounded by a jacketed space as in the average automobile engine. In former years, water was used exclusively as a coolant, but now, various liquids have been devised, which have more favorable cooling characteristics. The liquid is circulated by a pump and cooled in a radiator.

Another type of aero engine which is very widely used is the air-cooled radial engine, so named because the cylinders are arranged radially around a central crankcase. In this engine there is only one throw to the crankshaft and one master connecting rod attaching to the crankpin. All the other connecting rods attach to the master.

The order of firing is alternate, as in the case of the nine-cylinder engine, 1,3,5,7,9,2,4,6,8. There must be an odd number of cylinders in such an engine, otherwise the same alternate cylinders would be firing all the time.

The cooling of this engine is brought about by the fins which surround the cylinders. It is a well-known characteristic of heat that it flows to the thinnest point. Thus, the heat travels to the cooling fins from where it is carried off by the air.

A special structure is built around such engines to cut down their air resistance in flight, yet provide enough air around the cylinders to carry off the heat. This is called the "cowling."

The liquid-cooled engine, however, may be more completely cowled to conform with the shape of the fuselage or wing than the radial engine, as it is not necessary to divert any air around the cylinders for cooling purposes.

So specialized is the manufacture of aero engines that it is not the practice to build them in the airplane factory. Most success has been achieved by those corporations whose only aim is the development of the aero engine.

The prospective apprentice must therefore decide beforehand what specific line he wishes to follow so as not to waste valuable time; then he must go ahead.

Either field presents unlimited opportunities for the boy who can combine ambition with a mechanical inclination.

CHAPTER 6

ELEMENTS OF BLUEPRINT READING

One of the most important phases of mechanical work of any kind is to convey the ideas of the designer to the craftsman in the shop, so that the mechanism or detail may be built correctly. Verbal instruction in this respect would be out of the question as it is too vague, too changeable and unofficial.

It is therefore necessary to adopt a plan which is more positive in its operation and which reduces the possibility of mistakes to a minimum. The designer draws his idea on a piece of transparent tracing paper, after which it is printed. This consists of passing the tracing, in contact with a piece of sensitized paper, through a machine which exposes, develops, and dries the print in much the same manner that a photograph is made. The pencil lines on the tracing are reproduced in white, the entire background being transformed to a bright blue.

This is known as the blueprint and is sent to the shop as a pictorial record of the size, shape, structure and material of the mechanism to be built. The blueprint paper itself is of a tough, hard-wearing texture, so that it will withstand the rigors of shop handling. The tracing is filed away so as to be available for future reference.

Title Block—Every blueprint in the aircraft factory has a title and a number. The latter is absolutely necessary for the filing of records. In the bottom, right corner of the tracing is printed the title block, and when properly filled in by the draftsman, it supplies all the necessary information about the drawing that is not given in the picture itself. Incorporated in the drawing number sometimes there is a code letter which signifies the over-all dimensions of the blueprint. In addition to this, the model designation of the airplane is noted, as are the signatures of the draftsman, the checker, the stress analyst, the chief draftsman and, in some cases, the customer's representative. Space is also provided for the dates of these signatures.

The permissible dimensional limits are also inserted here, of which more mention will be made presently. If the drawing is of one detail only, the material and the heat-treat specification will be noted in the title block. The apprentice must learn to regard the material and heat-treatment as very important, because in most cases the heat-treatment must take place at a very definite stage in the manufacture of the piece. Any deviation from this usually results in scrap, due to the warping of the metal in heat-treatment. Provision must be made for a subsequent machining or forming operation for straightening. The draftsman must also insert the scale size of the drawing, which he maintains throughout. This helps to give the mechanic a visual picture of the size of the object.

A very important part of the title block is that which carries the drawing-change notices. Especially in the aircraft industry, where design changes almost constantly, it is necessary to record all drawing changes. If any blueprint is not up-to-date according to the latest engineering information, it should be disposed of immediately, otherwise more scrap will be the result.

Types of Drawings—There are three distinct types of drawing: perspective, isometric, and orthographic. In the perspective drawing, the object is shown as it would appear in a photograph. This means that as the distance from the eye (or camera lens) increases, the size of the object seems to taper off. In the isometric drawing, the length, breadth, and thickness of the object are shown in one view, but the tapering off due to distance is ignored. The orthographic system is the accepted practice for mechanical drawing in

America. In this type of drawing, the object is viewed from three different points, so that three distinct pictures are shown. These views or projections, as they are called, do not show the object as the eye would see it, but they give the outline as it is known to exist (Fig. 16).

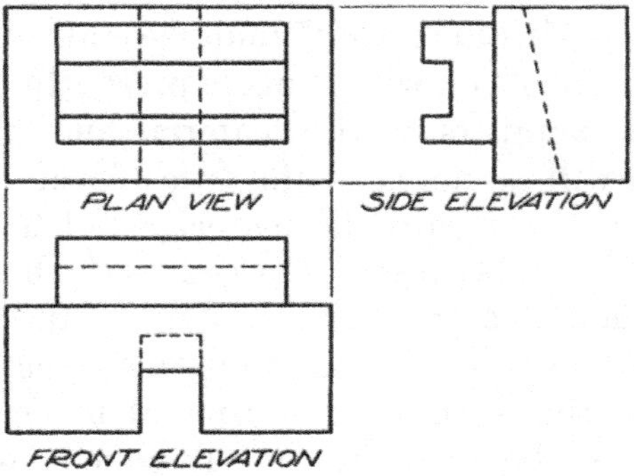

Fig. 16

Views are known as top or plan, front, and side views, and are so named from the position of the eye in relation to the object. In some cases, due to the shape of the piece, a bottom view may be used in place of the top, and the side view may be either right or left. Sometimes it will be deemed necessary to show both right and left views on the same drawing. Some objects can be readily portrayed by two views only, and in the case of a piece of tubing or round bar these can be clearly shown in one view, if one of the dimensions is indicated as a diameter, by the symbol "dia." Another method of portraying a cylindrical object is shown in the sample drawing, Fig. 17.

Besides showing the shape of the object, the blueprint must give all the information about its size. The various sizes are known as dimensions, and the apprentice must learn how to interpret the blueprint in this respect. The sample drawing shows clearly that the length of the round tube is 8¼ in. The extent of the dimension just mentioned is indicated by two arrow heads which touch two extension lines coming from the ends of the tube. It will be noted that the extension lines do not actually touch the tube. The actual dimension figures always should be placed at right angles to the dimension lines.

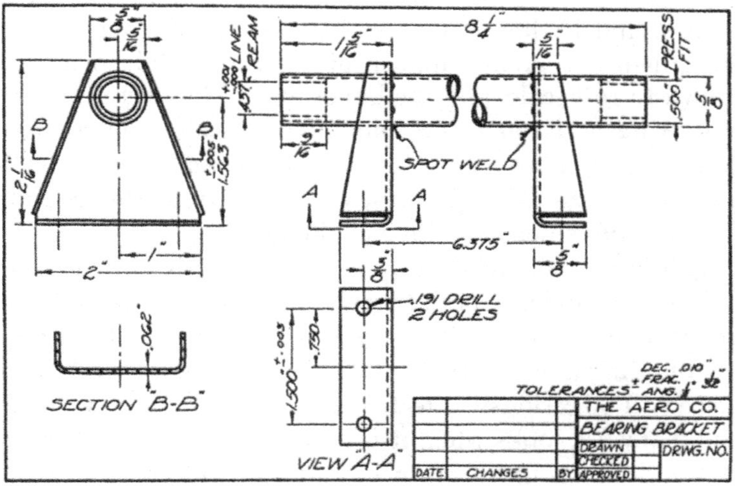

FIG. 17

Dimension Limits—Due to the fact that mechanical perfection is impossible, all blueprint dimensions must allow the mechanic certain permissible limits of error. When a dimension does not have to be very accurate, it may be expressed in inches and fractions. When accuracy is desired, the dimension must be expressed as a three-place decimal. Just above the title of the drawing may be seen the note that states the limits for all the dimensions, unless otherwise specified.

The plus and minus signs in this note mean that any nominal dimension may vary by the amount stated. For example, the length of the tube in the sample drawing, being a fractional dimension, would be acceptable if it were made to a length of 8 9/32 inches, or to only 8 7/32 inches. When it is necessary to obtain a greater degree of accuracy than is stated in this note, the limits are placed with the particular dimension involved. (See sample drawing, 1.500 dimension.)

Although every blueprint should be made to scale, the apprentice must never measure the drawing to obtain any dimension. Such information must be found by mathematics, and if this is impossible, then the draftsman must be called. A great number of aircraft detail prints are drawn to full scale, and the larger assemblies are reduced to half scale and sometimes one-quarter scale. There is no actual limit, however, to the amount of scale reduction for a blueprint, as this is governed by the size of the assembly and the convenient handling of the print.

The letter "R" after a dimension signifies that it is the radius of a circle. Special attention should be given by the apprentice to any radius governing the amount of metal surrounding a hole. If this is undersize, the strength of the piece will be seriously affected, and the apprentice will learn how his work looks when it is adorned with a rejection tag.

A hole on a flat surface is indicated by cross lines, the point of intersection being the center of the hole, around which a circle is drawn. In the case of a number of holes of the same diameter in the same vicinity, a note will appear on the drawing stating the diameter of the finished holes, the type of hole (drilled or reamed), and the number of holes required. From this note an arrow will point to one of the holes. Great care must be exercised by the apprentice in selecting his drills and reamers. An oversize hole is a job scrapped.

The initials "O.D." or "I.D." accompanying a dimension mean outside diameter or inside diameter.

Ordinates—In aircraft blueprints the apprentice will often find curves that do not remain equidistant from any one given point, as is the case with a circle. These are termed contours and are dimensioned by means of ordinates. A series of parallel lines, usually two inches apart, are drawn at right angles to a reference line, the curve passing through each of the parallel lines. The dimension from the reference line to the intersection point on each parallel line is then inserted. The joining of all the intersection points produces the curve.

Sectional Views—Due to the fact that in some intricate shapes it is impossible to visualize the structure from the three or four conventional views, another type of view has to be utilized. This is termed the sectional view. By theoretically cutting the object through a plane designated by the section line, the hidden structure is exposed and shown in the sectional view, which is drawn in any convenient location on the blueprint. In the sample drawing, Fig. 17, the sectional view is identified with the section line by the letters B-B. Any such additional views must carry other identifying letters. It is very important that the sectional view be shown in the direction indicated by the arrows at the extremities of the section line. The section that has been cut must be identified by diagonal lines or cross hatching, as it is called. All hidden edges or surfaces, such as the inner wall of a piece of tubing, are shown by dotted lines.

The Bubble Diagram—Although the foregoing remarks are applicable to all blueprints, those that the apprentice will encounter in the aircraft factory are divided into three groups: detail prints, assembly prints, and installation prints. The detail print depicts only one piece; the assembly print shows two or more individual pieces attached together. The method of attachment may be bolting, riveting, welding, soldering, or threading. It is also important that each detail on the assembly must be identified. If it is possible to dimension completely any detail on the assembly print, it is allotted a dash number following the assembly number. This effects a great saving in paper, because such details are manufactured without the use of detail blueprints. If, however, on the assembly any detail has its own print, it is identified by that print number. Each identifying number on an assembly is enclosed in a circle, usually termed a bubble. In order to help the mechanic in locating all the details on the assembly print, the bubbles are all shown with their detail numbers on the right-hand side of the blueprint, besides being appended to their respective details on the drawing itself.

Sometimes the draftsman divides the assembly into smaller subassemblies in order to facilitate production. In such cases each subassembly receives a dash number, and in the bubble diagram all details are grouped under their subassembly dash numbers. In this way it is comparatively easy to check up for missing details, especially on large assembly drawings. It also facilitates the setting up of a storage and distribution system, by properly identifying each piece and assembly.

Installation drawings, as the name implies, are made for the purpose of showing the mechanic where and how the various assemblies are located on the airplane.

An important fact that the apprentice must remember is that not all the information on the blueprint is in pictorial form. Every written note must be studied, because they sometimes vitally affect the whole drawing. For instance, if a drawing note says that certain holes must match some other piece and the other piece is governed by limits of plus or minus 1/32 in., then care must be exercised so that both pieces will match within close tolerances.

Many parts of the airplane are right and left hand, but in such cases the blueprint shows only one hand. The apprentice must therefore learn how to constitute an opposite hand from the view shown. Generally speaking, when two opposites are placed back to back they match one another.

The Drafting Apprentice—The drafting-room apprentice at the outset of his career is usually put to work transferring from drawing paper to tracing paper the drawings of the more experienced men. To do this he spreads the original drawing on a drawing board and fastens it down with thumbtacks. He then places a sheet of tracing paper (transparent) over the original, and fastens this also with thumbtacks. With the lines of the original showing through the tracing paper he retraces the complete drawing. It is essential that he should be very neat and clean in doing this because a smudgy, ragged drawing will not be accepted.

The alert apprentice may gain valuable experience from this phase of his work, because it affords him an opportunity of studying drafting as practiced by older and more experienced men.

His next assignment may be to draw details. Although he will be under the supervision of a group leader, he will gradually develop his own initiative in designing. His success as a detailer may be followed by his promotion to the layout group, who are responsible for positioning the details of the airplane.

His further promotion necessarily must be governed by his own special ability, and may lead him into aerodynamic, stress analysis or liaison work. The drafting apprentice, like the shop man, should obtain good tools, and he should report for work with the following:

Drafting Apprentice's Tool Kit—

1. Set of instruments (bow compass, large compass, dividers, ink ruling pen)
2. 12-in. scale (full and half scale)
3. 60-deg. triangle
4. 45-deg. triangle
5. French curve
6. Protractor
7. Erasing shield
8. Pen holder
9. Assortment of pen points

Large scales, special curves, splines and drafting materials are usually provided by the company.

CHAPTER 7

MACHINE TOOLS IN AIRCRAFT

For many years the mechanical engineering industry was divided into only two branches. First was the scientific branch which embraced that great field of advanced thought that has accompanied this industry down through the ages. Second was the productive branch which concerned itself with the more material task of fabricating the finished article.

Origin of Tool Engineering—Due to the ever-increasing demand for things mechanical, to the scarcity of skilled workmen and to the increasingly intricate nature of mechanical assemblies, the productive branch of the industry began to lag behind. Civilization demanded, but its demands could not be met. The result was that there grew up in the industry another branch called tool engineering, whose main object was to speed up production. So vital has tool engineering become that it now dominates the entire productive field. It has endowed the worker with a pair of mechanical hands that work a hundred times faster than his own; it has given him an eye for accuracy that he never dreamed of before. It is very natural, then, that aircraft manufacturers should rely on the forces of tool engineering in order to cope with the enormous demand for airplanes and also to enable them to produce at a profit.

Because this industry is comparatively new and skilled workmen are extremely scarce, the tool engineer is faced with the job of devising means and processes so that an uninterrupted flow of production will exist, while still maintaining the necessary high degree of accuracy in each detail. Quite a gigantic task, young man, and one that calls for a very extensive knowledge. Keep your eyes open and observe the use of high-production tools all around you.

Classification of Machine Tools—It must be remembered that the word "tool" is a very general term, and it may belong in any one of four groups.

A. Machine tools for removing metal in the form of chips. They are discussed briefly in this chapter.

B. Metal fabricating tools for forming sheet metal to certain shapes. These are discussed in Chapter 9.

C. Subsidiary production tools such as punches, dies, jigs and fixtures. They are used in conjunction with class A and B tools, and are made by the toolmaker. They are discussed in this chapter and also in Chapters 8 and 9.

D. Hand tools, which include the personal tool kit of the mechanic, and all precision measuring tools such as are described in Chapter 8.

Machine-shop Apprentice's Tool Kit—The class D tools that the apprentice should acquire in order to make a good start are:

1. Ball-peen hammer (8 oz.)
2. 6-in. flexible scale
3. 12-in. combination square with protractor
4. Scriber
5. Screwdriver
6. Center punch
7. Pliers (broadnose)
8. Adjustable wrenches (1 large and 1 small)
9. Hacksaw
10. Cold chisel

As the apprentice progresses in his work, he should add to this according to his needs.

Machine-shop Practice

In the machine shop may be seen some very interesting products of the ingenuity of the tool engineer. Let's listen to the talk that goes on in this department between the foreman and the operatives. To the apprentice it all seems very unreal, as if he had entered a new world where a new language is spoken. They talk of drill jigs. The apprentice knows that a "jig" is some sort of a Gaelic dance, but the connection is too obscure for him to trace. Actually, there is no connection. They talk of vise jaws, and the words have a sinister sound to the boy just out of high school. He overhears someone mention jig borer and imagines some terrible, winged agricultural pest. He reads a chart which depicts revolutions per minute and surface speeds in feet per minute, and for all he knows he may be looking at some abstract astronomical theory. But do not be alarmed, young man; a study of these things will convince you that with good common sense you can easily master the problems that arise in the machine shop. The basic function of this department is to remove metal, be it from solid bar, casting, or forging, in order to obtain a shape that adheres to the dimensional limits of the engineering drawing.

The entire industrial structure of the world revolves around, and is supported by, the machine tool, and almost every phase of our physical life is influenced by it. These are very broad claims, but they are justified when we consider that every automobile, electric refrigerator, home heating plant and radio owes its present stage of development to the machine tool, and that the present worldwide systems of transportation, manufacturing, communication and agriculture are directly dependent on that which is created by the machine tool. The versatility of the machine tool is amazing, its products ranging from such jobs as the giant turbines of the liner *Queen Mary* to the gears of a tiny wrist watch.

Methods of Machining—There are five basic methods by which the machine tool may remove metal: milling, drilling, turning, planing, and grinding. Each of these methods may be accomplished by several different kinds of machines, each being developed for a specialized type of its own basic method of machining. Thus we have horizontal, vertical, universal, hand and special extrusion milling machines. We have radial, multi-spindle, and single spindle drilling machines, and cylindrical, surface, centerless and thread grinding machines. The essential thing for you, the apprentice, is to know something of the basic operations.

Certain subsidiary devices are often used in conjunction with the machine tools in order to produce specific parts, and these are called jigs and fixtures.

1. MILLING

Milling Machine—Let us study the action of the milling machine shown in Fig. 18. To "mill" a piece of work is to bring it in contact with a revolving cutter, which has a number of cutting edges around its circumference and which removes the metal from its path in the form of chips. The apprentice may wonder how it is possible actually to cut into a piece of steel with another piece of steel in the form of the cutter. He knows that it would be impossible to whittle a piece of wood with a wooden knife. But the tool engineer knows enough about the science of metals, or metallurgy, to know how to select the steels from which cutters are made, and how to heat treat them in order to make them many times harder than the steels which they are expected to cut. However, we shall discuss tool steels and cutters a little later, so for the present let us continue our milling operation with the assurance that our cutters are in perfect order. The spindle which carries the cutter of the modern milling machine is motor driven through a transmission box, which enables the operator to use his discretion as to cutter speed while taking into account the diameter of the cutter and the composition of the metal which is being cut. The table which holds the work is capable of movement and can be set to any desired distance from the tips of the cutter teeth, thus determining the depth of the cut. It can also be moved in any horizontal plane, thus giving the cut length and breadth as well as depth.

Courtesy Brown & Sharpe Mfg. Co.

FIG. 18—OPERATING MILLING MACHINE IN A DETROIT HIGH SCHOOL

1. Revolving cutter
2. Indexing fixture
3. Table
4. Handwheel for vertical movement of table
5. Motor
6. Handles for regulating spindle speed
7. Reversing handle for spindle
8. Handle for regulating feed
9. Spindle
10. High- and low-speed lever for feed

Speeds—The peripheral speed of the cutter is very important. If excessive, the friction between the cutter and the work will generate so much heat that the carbon content in the cutter will be destroyed and the cutting edges will break down. In order to operate at the highest possible speed, however, it is usual to direct a stream of cutting oil on the tool, so that the heat may be dissipated and the chips cleared away. The feed or horizontal movement of the table is also very important. If this is excessive, it also will generate too much heat, as in the case of too much cutter speed and with the same result, as well as causing undue wear on the feed screws of the machine itself.

Fixtures—In most cases it is not possible to clamp the work directly to the table, due to the shape of the former. It is, therefore, necessary for the tool engineer to design a fixture that will hold the job rigidly, in the proper relation to the cutter, and that will provide easy loading and unloading. Sometimes it is necessary to perform two or more operations on one piece of work, and this can sometimes be accomplished without removing the piece from the fixture. A simple form of such a fixture is shown in Fig. 19. The

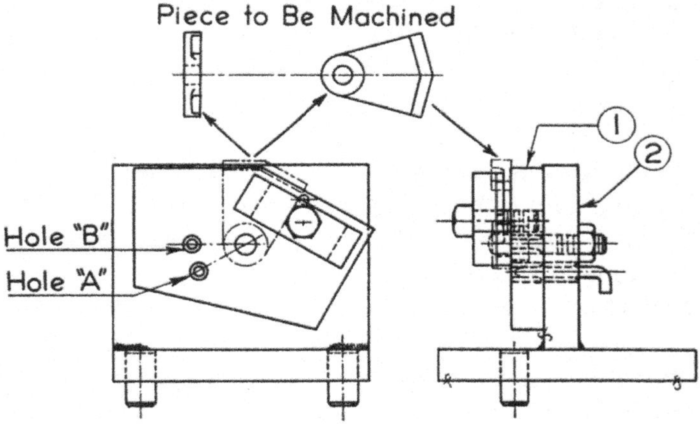

FIG. 19—PIECE IN CLAMP

work is clamped to detail 1; a pin is inserted in the hole A, which is concentric with, and of the same size as, a hole through detail 2; and the pivot nut is tightened. The first surface to be machined is then in position. In order to bring the second surface to machining position, the pivot nut is loosened, the pin is removed, and detail 1 is rotated until hole B is directly over the hole in detail 2. The pin is again inserted, the pivot nut is tightened, and the second surface to be machined is in position. A fixture of this type is said to be an "indexing fixture."

Spiral Milling—It is interesting to note that spiral milling is accomplished by rotating the work as well as the cutter. A good example of this operation is the milling of an ordinary twist drill. The round blank is placed between the centers of an indexing fixture and parallel with the tee slots of the milling-machine table. The latter is then set to the proper helix angle and the indexing fixture is geared to the feed screw of the milling machine. The ratio of the gears selected must be such that, when the work has made one complete revolution, the table has moved a distance equal to the desired lead of the spiral to be cut. The number of flutes to be cut is gauged by the manual operation of the indexing head. It is in this way that spiral reamers, end mills, and other similar cutting tools are produced.

2. DRILLING

Continuing our tour of the machine shop, we must pause with that group of operators who possess a strange faculty. With the removal of metal, the object of their work becomes larger. You've guessed it, they drill holes.

Drill Press—The importance of the work accomplished by the drill-press operators cannot be overestimated, as it is impossible to get a good engineering job of any kind, if the necessary holes are not of good quality. In the first place, every hole must be in the proper location in relation to the other holes or surfaces. The dimensions which govern the location of holes are usually given with limits which specify the amount of permissible error. Generally speaking, these limits range from 1/32 in., when the locating dimensions are in fractional form, to plus or minus .0005 in., where accuracy is desired. Secondly, a good hole must measure up to the specified diameter, within the limits of error as in the case of location, and must be free from taper—except, of course, where a tapered hole is called for. A good hole must also be perfectly round, and when a reamed hole is specified on the blueprint, the wall must be smooth. This is of prime importance. If a hole is full of ridges, the crests will soon be worn down by the bolt or pin, when the load is applied in service. When this happens, the hole becomes "sloppy," and additional wear on both bolt and fitting takes place rapidly. The ideal condition is a hole with an unbroken wall, and as near size to size with the bolt as possible, so that the latter will be restricted from any movement whatsoever.

In order to help the drill-press operator to accomplish these things the tool engineer has devised many appliances. The modern drill press consists of a motor to supply power, a table for supporting the work, and a vertical spindle which rotates on its own axis and is capable of up and down movement along its own axis. This spindle carries a chuck at its lower end for gripping the drill. The pressure of the drill against the work is applied by means of a hand lever which acts through a rack and pinion. The speed of the drill must be decreased as the diameter increases in order to prevent excessive peripheral speed which would break down the cutting edges of the drill. Cone pulleys are introduced to the driving mechanism for this purpose. The mechanical principle of the cone pulley is shown in Fig. 20. As in the case of the milling machine, cutting oil may be applied for the same purposes.

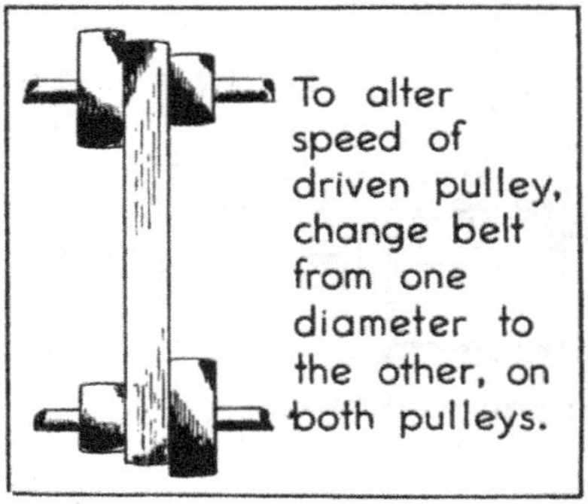

Fig. 20

Radial Drill Press—Another more elaborate type of drilling machine is known as the radial drill press. A large arm which swings radially from a vertical post carries the mechanism for both rotating and applying the pressure to the drill. This unit is also capable of moving outward along this arm, allowing the drill to operate in any location within the sector covered by the swing of the arm. For this reason the radial drill press is more versatile than the ordinary drill press, as it is much easier to move the drill to the desired hole location than to move the job itself in relation to the drill, especially on large jobs where numerous holes have to be drilled.

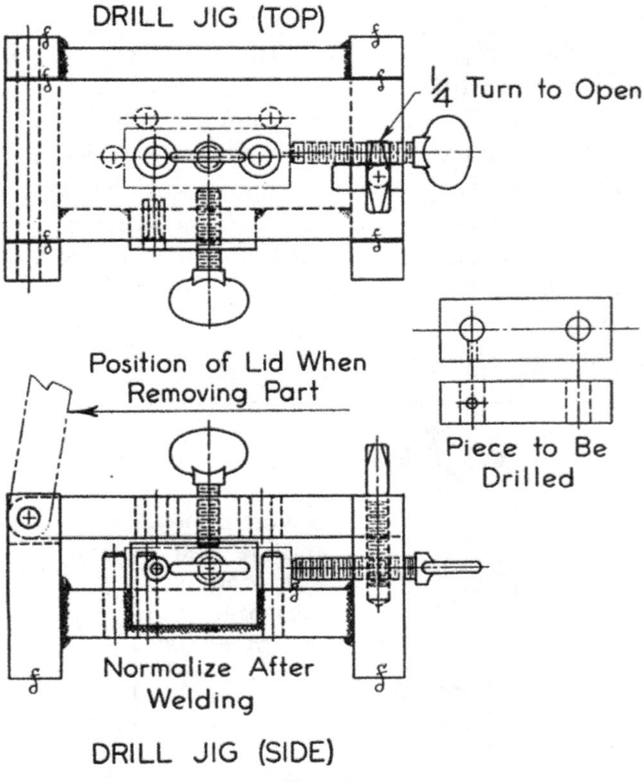

FIG. 21

Drill Jigs—The modern drill press, however, be it radial or conventional, is not in itself capable of drilling enough holes in a given time to satisfy modern requirements. On a production job it would take a vast amount of time to mark off each hole accurately and to indicate the center of each hole with a center punch indentation, so that the drill would start cutting in the proper location. Absolute accuracy would also be impossible by this method. The tool engineer knows that the saving of time means the saving of money, and that the saving of money means more profit to the company. This in turn means more to re-invest in plant expansion, better equipment, and more experimentation, bringing more work and pay to the drill-press operator. He therefore designs drill jigs to speed up production.

The fundamental purpose of any drill jig is to hold the part firmly in the proper relationship to the drill so that the hole will assume the correct location. The jig must also be so constructed that the operator can load and unload in the shortest possible time. Drill guides must be provided and are known as bushings. These are made of hardened tool steel, so that they will not wear readily from the action of the drill. In Fig. 21 is shown a drill jig and also the part to be drilled. You will notice that the jig is of welded-steel construction, being normalized, then machined at the vital points.

3. TURNING

Engine Lathe—One of the most interesting of all machine tools is the engine lathe. The primary function of this machine is to produce round work. When a suitable cutting tool is brought into contact with the work, which rotates on its own axis, between centers, a round shape is developed. The lathe consists of a bed which carries the headstock, the tailstock and the saddle. Protruding from the headstock is the work-supporting center, another being located in the tailstock. Power is supplied by a motor, through a selective gear box to the driving spindle. The end of this spindle is threaded to receive a faceplate or chuck, depending on the type of work on hand. The saddle on which is mounted the tool post is capable of movement along the bed, and the tool post itself can be moved in a transverse direction so that the diameter and the length of the work can be determined at the will of the operator. In producing a round spindle, the operation is called "turning."

Boring—The lathe, however, is also capable of that operation called "boring." As the apprentice may imagine, boring has to do with holes, but it may be distinguished from drilling inasmuch as the boring tool does not cut a hole from the blank, but only enlarges an existing hole. The conventional boring tool consists of a bar which carries at one end a tool steel cutter, similar to the turning tool. As the boring proceeds, the bar enters the hole, and the cutter is thus supported all the way to the ultimate depth of the hole. With boring, it is possible to change accurately the center location of a hole, the eccentricity of the original hole being evident in the constantly changing depth of cut during each revolution of the work. On the other hand, a twist drill will tend to follow concentrically the original hole. In lathe operation, large holes are usually started with a twist drill and finished to an accurate diameter with a boring tool. It is also possible to cut both external and internal threads on the engine lathe. The cutting tool is ground to fit a thread gauge, and is set so that the angle of each side of the thread will be equal, in relation to the axis of the work. The lead screw of the lathe is then geared to the spindle so that the saddle of the machine will move a distance equal to the pitch of the thread, during each revolution of the work. Right-hand threads are cut with the tool moving towards the headstock, and in the case of left-hand threads, the tool moves towards the tailstock while cutting.

Turret Lathe—Also included in the basic art of turning are the "turret lathe" and the "automatic lathe," both designed for high production. Fig. 22 shows the turret lathe. As can be observed, a battery of cutting tools is contained in a hexagonal turret, which, as it is revolved by the operator, brings each tool into operation in the proper sequence.

Courtesy Warner and Swasey Co.

FIG. 22—HEAVY-DUTY TURRET LATHE, MACHINING AIRPLANE WHEEL

1. Hexagonal turret
2. Boring bar
3. The work being machined
4. The coolant
5. Handwheel for adjusting tool

Two additional tool posts, one at the front of the machine and one at the rear, are also available for any operation such as forming the outside diameter of the work to a particular shape, in which case a special form tool is prepared. or for cutting off the completed work from the bar. In this way it is possible to complete such operations as centering, drilling, boring, reaming, turning, chamfering, forming and cutting off, all at one setting.

Automatic Lathe—As the name implies, the automatic lathe has been designed to operate without the guidance of the human touch. The movements of the various tools are controlled mechanically, and after the machine has been set up by a skilled operator to run any particular job, it will persist in doing so until the cutting edges of the tools are

Courtesy Cleveland Automatic Machine Co.

Fig. 23—Cleveland "Automatic" Making Cam-bearing Nuts, for Airplane Propellers

1. Tool holder on cross slide
2. Cross slide
3. Motor
4. Tool turret
5. Cam for feeding material

dulled or the supply of raw stock is exhausted. In Fig. 23 is illustrated a large automatic lathe, turning out cam-bearing nuts for airplane propellers, while in Fig. 24 can be seen the tool set up for the same pieces. Note that six pieces are being cut off at one time.

4. PLANING

Planer—A glance at Fig. 25 will convince the apprentice that the planer is really the big brother of the average machine tool. Compare the relative sizes of the man and the machine. The planer consists of a table which reciprocates horizontally along two supporting "ways." A cross-rail extends over the table and carries the cutting tool. The work is clamped to the table and reciprocates with it, and

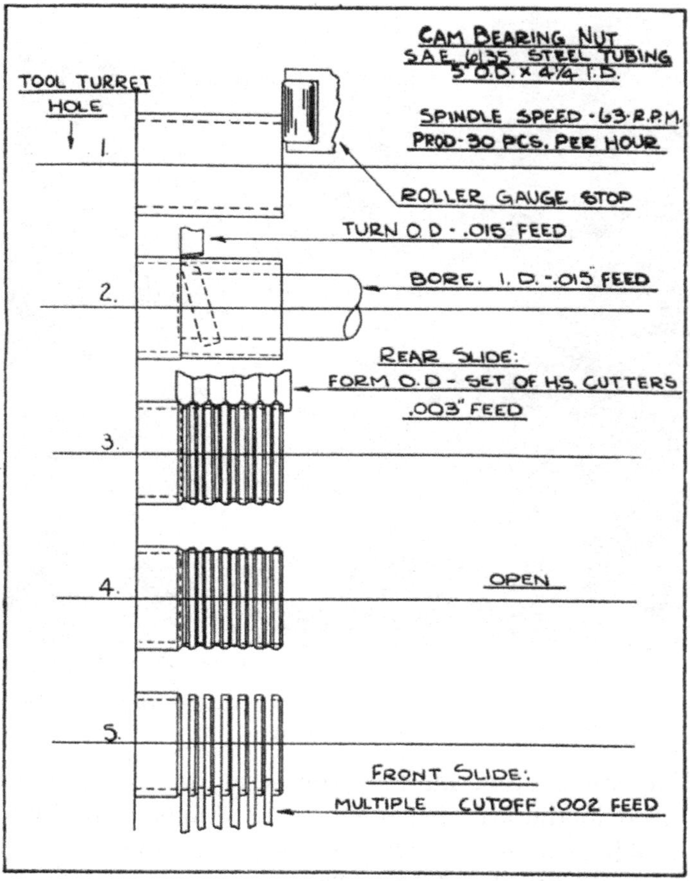

Courtesy Cleveland Automatic Machine Co.

FIG. 24—TOOL SET-UP FOR JOB SHOWN IN FIG. 23

the cutting tool can be moved up or down in order to come in contact with the work, producing the proper depth of cut.

After each cut, which is produced in a straight line along the length of the work, the tool can be subjected to a mechanical sidewise movement, a succession of which causes the planed portion of the work to assume width as well as length. Fig. 26 shows the actual planer whose "ways" were being planed in Fig. 25.

Courtesy G. A. Gray Company

Fig. 25—Interesting Example of Machine Tool Reproducing Itself
1. Planer "ways" being planed

Shaper—Another machine the action of which must be included in the art of planing is the shaper. In this case the tool instead of the work reciprocates, producing straight-line cuts as with the planer. The feed, however, is produced by the sidewise movement of the work. The shaper has been developed to take care of work that would be too small for economical production on the planer.

5. GRINDING

We now come to the last and most recently developed art of metal machining, that of grinding. Although in prehistoric times, our primitive forebears made use of sand, flint, and naturally sharp rocks to whet their hunting and warring implements, it was only in recent years that our present sys-

Courtesy G. A. Gray Co.

FIG. 26—THE COMPLETED PLANER

1. Table
2. Actual "ways" shown being planed in Fig. 25
3. Cross rail
4. Tool holders

tem of grinding was developed. Among the five basic arts of machining, grinding assumes the responsibility of maintaining accuracy in high production. It takes up the task where the other machine tools fail entirely, because it is able to cut steel that has already been hardened. In modern grinding practice, two principal abrasive materials are used, and both are entirely artificial. The first, silicon carbide, is the result of a chemical union, under heat, of carbon and silicon that produces a crystal that is very hard but is slightly brittle in nature. The other, aluminum oxide, is not quite so hard but much tougher than the silicon carbide. With these crystals, graded to various uniform sizes and suitably bonded into wheels, it is possible to take care of all the grinding problems of industry, from the grinding of leather to the polishing of precious stones.

There are two distinct classes of grinding that should be brought to the attention of the apprentice. One is production grinding; the other is cutter grinding. The former is the actual production of industrial parts which have been selected for grinding because of their required accuracy or the extreme hardness of their composition. The latter is the process by which the hardened cutters for milling, drilling, turning, and planing are given that special form of cutting edge and clearance angle that will enable them to remove metal in the most efficient manner.

A. Production Grinding—In the aircraft factory the apprentice will find the production grinding divided into five different classes: cylindrical (external and internal), centerless grinding, surface grinding, thread grinding and hand grinding.

Cylindrical Grinding—In the external grinder, the work is usually held between centers, as in the lathe, and is revolved at approximately 25 surface feet per minute. In contact with the work, the grinding wheel must revolve at approximately 6000 surface feet per minute. As in all other grinders, the bearings of this machine must be exceptionally rigid and free of play, in order to produce an unblemished ground surface. In operation, a stream of fluid, usually termed the "coolant," is directed on the wheel so as to clear away the ground particles and keep the work cool. The feed in this machine is produced by moving the work sideways in relation to the wheel, and on the modern grinder this is accomplished hydraulically, so as to get away from the chatter which sometimes accompanies the mechanical feed. In conjunction with the external grinder can be used the automatic indicator gauge, which reveals to the operator at all times the diameter of the work just ground. The dial on this indicator is graduated in tenths of a thousandth of an inch.

The internal grinder is very similar in operation to the external grinder, except that internal diameters are ground.

Centerless Grinding—In the centerless grinder a different principle is employed. Instead of being held between centers, the work is supported on a rest bar, between a regulating and a grinding wheel. The action of the grinding wheel forces the work against the rest bar, while the regulating wheel imparts its own peripheral speed to the work. In the case of plain bars, it is possible to feed them through the machine in quick succession by inclining the regulating wheel at an angle so that it pulls the work lengthwise, as well as causing it to rotate about its axis. When irregularly shaped pieces have to be ground, the in-feed method must be used instead of the through-feed. In this case, the grinding wheel must be dressed to the required shape and the regulating wheel set, so that no lengthwise movement of the work occurs.

Thread Grinding—Due to the ever-increasing demand for high-tensile and wear-resisting screw threads, the thread-grinding machine has been developed. By grinding the threads from the already hardened bar, the distortion due to the heat-treatment is eliminated, and the finish is of a very high quality.

Surface Grinding—It is also possible to grind flat work as well as round. This is done in the surface grinder. One type of this machine employs a conventional flat, round wheel, the axis of which is horizontal, and being capable of up and down movement, the wheel can be brought into contact with the work. The handwheel which operates this movement is graduated in thousandths of an inch. The work is held to the table by means of magnetism, so that the surface to be ground will not be obstructed by clamps. The table reciprocates in line with the motion of the wheel, the length of stroke being adjustable to suit the length of the job. A transverse feed allows the cut to cover an area within the capacity of the machine. The outer edge of the wheel comes in contact with the work, and a steady stream of coolant is used to prevent excessive heating and to wash away the chips. As the spindle of the wheel cannot be moved from the horizontal, all angular surfaces must be ground by either tilting the work or dressing the face of the wheel to suit.

In another type of surface grinder, suitable for much larger work, the spindle is vertical, and the wheel is built up of segments or "bricks," as they are called. This wheel is located at the bottom end of the spindle and has a butt contact with the work. Being of slightly larger diameter than the width of the table, the entire surface of the work is ground at one pass. In this machine also the clamping is magnetic.

Hand Grinding—In the bench and floor types of hand grinders, the work is supported by the rest in front of the wheel, and the pressure is applied by hand, at the discretion of the operator.

B. Cutter Grinding—Cutter grinding is a highly specialized branch of grinding practice, and it may be of interest to the apprentice to know something of this subject. As the hardening of tool steels for cutters requires temperatures from 1400° F. to 2200° F. at which heat the delicate cutting edges would break down, it is absolutely necessary to grind after hardening.

Grinding Lathe and Planer Tools—Lathe and planer tools are usually ground by hand, the skill of the operator being responsible for producing the correct shape for the job on hand. Certain fundamental laws, however, must be observed. As viewed from the top (Fig. 27), the lathe tool must have a certain shape, according to the work it is required to do. There must also be a certain amount of front and side clearance, starting immediately below the cutting edge, and a definite slope must be given to that surface of the tool which carries the chip immediately after it has been severed. The lathe operator who may be a golf enthusiast would term this as "follow through." The importance of this cannot be overestimated, as it has a direct effect on the finish of the machined surface. This "rake," as it is properly called, should be such that the chip is conducted away with a minimum amount of power load on the ma-

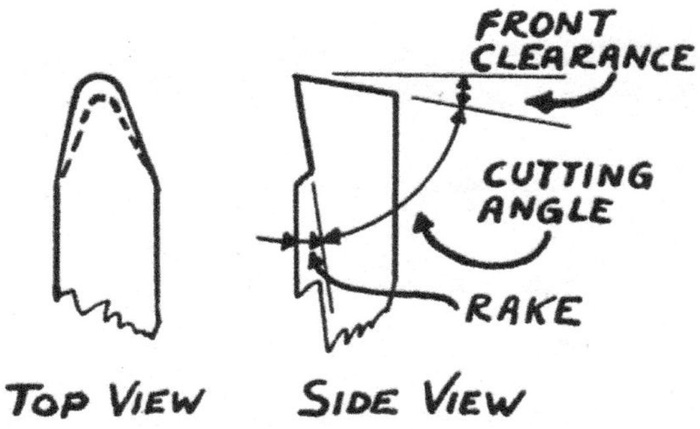

FIG. 27

chine. If incorrect, the chip, instead of being cut, will be literally torn away, and this results in an undesirable finish, high power costs, and undue wear on the vital parts of the machine itself.

The amount of cutting angle, which is 90 deg. — (rake + clearance) should be from 60 to 75 deg. for cast iron, and from 40 to 50 deg. for steel. If the cutting angle is too small, the tool will lack enough body to dissipate the heat, and enough strength to withstand the cutting pressure. Short tool life will be the result. Planer tools are governed by the same general principles as are lathe tools.

Grinding Twist Drills—The modern twist drill is a very efficient cutting tool and, due to its capacity for operating under very adverse conditions, it does not as a rule receive the attention that it should. To assure effective production, however, three important features in the grinding of twist drills must be maintained. First, the cutting edges of a twist drill are called "lips," and each lip must be exactly the same length. If this is not the case, the long lip, while cutting, will develop a longer radius than the nominal diameter of the drills calls for, which results in an oversize hole. Second, the angle between each lip and the axis of the drill must be equal. If not, one lip will do all the cutting and will pull the drill to one side, causing it to rotate about a

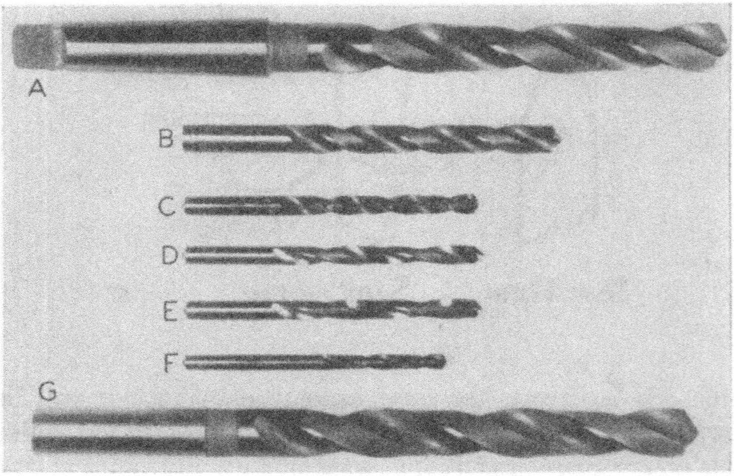

Courtesy Morse Twist Drill & Machine Co.

FIG. 28

A. Morse taper drill
B. Jobbers drill
C. Heavy-duty drill
D. "Bakelite" drill
E. "Brass" drill
F. Cotter-pin drill
G. Straight shank taper-length drill

point other than its own axis. An oversize and uneven hole will again be the result. Third, it is very important that the correct clearance be maintained from the cutting edge, backwards. If not enough clearance is provided, the area behind the lip will rub against the bottom of the hole and will prevent the cutting edge from coming in contact with the work. Burned up drills are the result of this condition, due to the excessive friction. If too much clearance is provided, the drill will have a tendency to gouge into the work.

very often resulting in drill breakage. It is possible to grind twist drills by hand, but for quick and accurate results, it is best to use the drill grinder, a machine specially designed for that purpose.

Grinding Milling Cutters—While it is the general rule to grind lathe and planer tools and the smaller twist drills by hand, it is absolutely essential that milling cutters and reamers be ground by machine. This is a very important feature in the efficient operation of the machine shop. The cutting edges of the milling cutter must be concentric with the arbor hole, otherwise a series of ridges on the finished surface of the work will result.

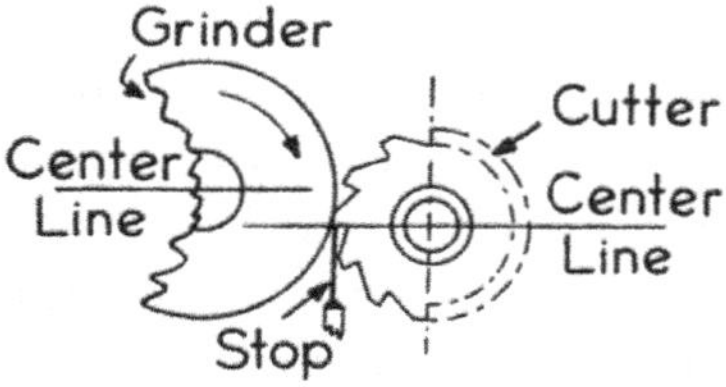

FIG. 29

FIG. 30

The cutting edges must also be kept very keen by frequent grinding, so that it will not be necessary to remove a large amount of metal to restore the edges.

There are two basic methods of grinding milling cutters. In Fig. 29 it will be seen that the center of the cutter is below that of the wheel. This produces what is known as the primary clearance, or "land." The action of the wheel against the cutter tends to keep the tooth pressing against the stop, but at the same time causes a burr to overhang the cutting edge. This must be removed by stoning. The burr may be reduced to a minimum by taking light cuts. Care must be taken not to draw the temper out of the tooth by overheating, because of too heavy a cut. The cutter is mounted on an arbor, which in turn is supported between centers on the table of the grinding machine. When the table is moved lengthwise therefore, the wheel will come in contact with the entire length of the tooth. If the cutter is spiral, it must be rotated on the arbor by hand while the table moves, so that the tooth will remain in contact with the wheel. Care must be taken to move the cutter clear of the wheel before any attempt is made to rotate to the next tooth.

In the case of the form cutter, the function of which is to impart a specific shape to the milled surface, a different procedure must be employed. The grinding must be done on the face of the tooth, so as to preserve the contour of the cutting edge (Fig. 30).

HEAT-TREATMENT OF STEEL

A few remarks now on the heat-treatment of steel may be of interest to the novice in the shop. He should bear in mind, however, that in all cases the hardening of steel should be left to those who understand fully the processes involved. The element in steel which enables it to become hard is carbon, and the change in its condition is caused by heating and subsequent cooling. The prescribed temperatures for different steels, however, must be accurately maintained. In general, the procedure is to heat the steel to what is known as the hardening heat (prescribed by the maker). It is then quenched in a cooling liquid, which may be brine, water, or various kinds of thin oils. The difference in the heat absorption of these liquids affects the hardness of the steel. For instance, a brine bath produces greater hardness than water, while oil results in a lesser degree of hardness. The temperature of the bath itself also affects the condition of the steel. After quenching, the steel is ready for "tempering," which means that the extreme hardness due to the quench must be drawn down by reheating to suit the particular application of the job on hand. This may be done by immersing in an oil bath and heating to a prescribed temperature, or by the color method, in which the hardness is gauged by the colors which appear on the surface of the steel. To do this, the steel must be clean so that the colors will show; then it may be placed on a red hot plate, the heat of which will cause the color changes. At first a faint yellow will appear, which will gradually change to a straw, light brown, dark brown, purple, blue, and dark blue. Each color indicates a certain degree of hardness, which decreases from the yellow to the dark blue. When the desired color (or hardness) has been reached, the steel must be quenched rapidly.

High-speed Steel—Most cutting tools are made from high-speed steels, which has reference to the high cutting speeds at which they will operate safely. It is the introduction of tungsten to the composition of high-speed steel which enables it to retain its hardness at very high cutting temperatures. The melting point of tungsten itself is a little over 6100° F. The cooling process may be done by placing the steel in a blast of dry air or quenching in an oil bath at room temperature, after which it is ready for use. This method, however, is suitable only in cases where the cutting tool has extremely well-supported cutting edges. Where a certain amount of toughness is required, the steel must be tempered by heating in an oil bath or furnace to a temperature ranging from 400° F. to 1000° F., according to the instructions of the steel manufacturer.

Annealing—Steel has also to be heat-treated in order to be softened properly, and this is called "annealing." This will remove all the unequal strains in the steel due to the rolling or forging during manufacture and will also make it more machinable. In general, the annealing of steel is brought about by bringing it to a prescribed temperature, slowly and uniformly, and allowing it to cool off, the softness or neutrality of the metal increasing with the length of the cooling period.

So that as little surface scale as possible will accumulate on the steel, it is advisable to place it in a box which has been scaled by a suitable plastic. It is the action of the oxygen in the atmosphere that causes scaling.

Case Hardening—Yet another well-known method of heat-treating steel is called "case hardening," which has some very distinct advantages. As the name implies, the external surface (or case) of the steel becomes very hard, while the core remains soft or tough, thus reducing the possibility of breakage due to brittleness. It is worthy also to consider that the low-carbon steels, which lend themselves to this heat-treatment, are comparatively inexpensive. The parts to be so treated must be placed in a box with some material containing carbon, such as charred leather or powdered bone. The box is then placed in the furnace, and the temperature brought up to approximately 1700° F. The carbon in the material starts to impregnate the steel after 1300° F. is reached, going deeper as the time under heat increases. If it is necessary to grind the finished piece, the carbon depth should be approximately 3/64 in. But in all cases this depth must be determined by studying the requirements of the job. The steel is then taken from the furnace and allowed to cool, after which it is reheated to about 1450° F. and immersed in a bath containing salt water, pure water, or oil. The salt water bath produces a harder case than does the pure water, but if a moderately hard case plus a tough core is required, the oil bath will produce this result.

HARDNESS TESTING

After a tool has been heat-treated as outlined, the most logical question for the mechanic to ask is: How hard is this tool? In order to eliminate guesswork and supply a definite answer to this question, several accepted methods of hardness testing have been developed.

Rockwell Machine—For hardened tool steel, the Rockwell testing machine is accurate. By bringing a small diamond cone in contact with the steel and applying a definite known pressure, a small indentation is made in the surface of the steel. A continuation of the load under additional known pressure gives a direct dial reading in Rockwell "C" units. For soft steels, the penetrator used is a hardened steel ball of 1/16-in. diameter, and the readings are in "B" units.

A Rockwell reading of 62 on the "C" scale denotes great hardness in tool steel, the toughness factor, however, being very low. As the toughness increases with the decrease of the hardness, it is necessary to obtain lower Rockwell readings in order to utilize the advantage of both hardness and toughness.

Brinell Machine—Another hardness-testing machine is the Brinell, in which a hardened steel ball is impressed into the surface of the metal. The diameter of the resulting impression is ascertained by a microscope. The load on the machine being known, the area of the impression gives the Brinell value.

Scleroscope—Hardness may also be tested by the scleroscope, a machine which measures the rebound of a small steel plug, after it is dropped on the surface of the test piece.

MOTION ECONOMY

The amount of control which the tool engineer exerts in the factory may be judged by the fact that he must outline the method of manufacture and sequence of operations of each individual job in the factory. In this connection the tool engineer must have a knowledge of what is known as "motion economy," so that he may design tools that will eliminate that millstone of waste motion that hangs around the neck of the aircraft industry. It has been said by a well-known industrial engineer that 30 per cent of the average worker's motions are waste, but this is considered a conservative estimate. Think of the saving, young man, if those motions were eliminated, and think of the opportunities that await the young man who possesses the inherent qualities of thrift, intelligence, and determination to eliminate them.

The application of motion economy does not mean making the operator work faster, because work done on the hurry-up plan is usually inferior. It is more essential that a comfortable working speed be maintained. By cutting out unnecessary footsteps and body movements, a higher rate of production results. By keeping up a comfortable bodily speed, the worker conserves his energy to last throughout the entire working day. For instance, when a small assembly has to be made, it is important that the detail parts be placed within arm's reach. The tools such as hammer, screwdriver, electric drill, rule, and the necessary punches should be placed in an orderly manner and also within reach, or better still a special rack should be provided.

If the nature of the job permits, the worker should have the option of either standing or sitting, because lengthy periods in one position produce sluggishness. When the small assemblies move along to the large-assembly lines, the tool engineer must see that proper handling equipment is designed. This may take the form of special trucks, hoist slings, roller conveyors, or even trunion devices for swinging complete fuselages in a lateral direction so that the assemblymen may apply the bottom skin while standing in a normal position.

It is very important that the tool engineering department should keep an accurate record of all the production tools (class C as mentioned earlier in this chapter) in the factory, so that when changes in design of the airplane occur, the tools affected may be changed accordingly and with as little delay as possible. It is usually necessary to keep a group of tool engineers for the sole purpose of seeing that all engineering changes are incorporated in the tools. So rapid is the development of aircraft that it is exceedingly doubtful if ever an airplane were built exactly to its original plans.

Machine-shop Tools—The following list classifies according to their function the various machine tools that the apprentice may encounter in the machine shop.

Milling
- Horizontal mill
- Vertical mill
- Hand mill
- Extrusion mill (having extremely long table and traverse for handling long extrusions)
- Thread mill

Drilling
- Single-spindle drill press
- Multi-spindle drill press
- Radial press

Turning
- Engine lathe
- Turret lathe
- Automatic lathe
- Horizontal boring mill
- Vertical boring mill

Planing
- Vertical shaper
- Horizontal shaper

Grinding
- External grinder
- Internal grinder
- Surface grinder
- Centerless grinder
- Thread grinder
- Floor hand grinder
- Bench hand grinder

CHAPTER 8

THE TOOLROOM

The tooling for aircraft obviously does not stop with the tool engineer. He creates ideas and puts them on paper, but it is the tool maker who reproduces those ideas in steel. In this standardized world, where the machine tool and its subsidiary devices are producing the implements of civilization, the tool maker is the master craftsman.

The apprentice should realize too that the toolroom is the most fertile recruiting ground for the tool-engineering department. So much practical knowledge is required by the tool engineer that it is impossible for him to acquire this in any technical institution. He must learn by doing, and so the toolroom becomes in part the Alma Mater of the average tool engineer. It is, therefore, much to the advantage of the toolroom apprentice to apply himself with an eye to the future. Hard work should be his daily companion, and the evening hours should see him bent over his text books, so as to build up his technical background.

For the first few months his task may not be very interesting and sometimes the apprentice becomes bored. He may think he is being slighted because he is asked to run errands or to clean up certain machines, but strict attention to these duties will tend to develop his self-control.

Toolroom Apprentice's Tool Kit—The toolroom apprentice may start to work with exactly the same list of tools as his friend in the machine shop, as described in the last chapter, but additions to his kit should be made from time to time.

Templates—Soon he will be advanced to the job of making templates or flat metal patterns. These are usually made from 1/16 in. thick galvanized sheet metal, and they occupy a very important place in tool work.

There are various kinds of templates, but in all cases their function is to indicate a certain profile. For instance, when

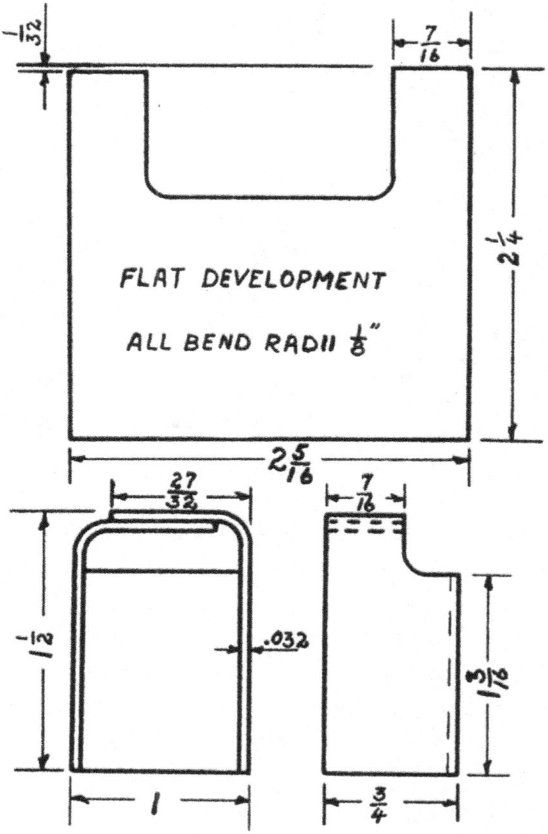

Fig. 31

a blanking die has to be made, it is first necessary to make a template, so that both punch and die can be scribed with exactly the same outline, and the shape may be inspected before the tool costs begin to mount.

Developed Templates—If the blank has to be formed to a certain shape later, the template must be made accordingly. This is known as a developed template, and, as it is usually made from a drawing of the finished piece, it is necessary for the tool maker to predict its shape before forming. Although this can be done with the use of a bend-allowance chart, it is essential that the apprentice should understand the theory that governs such calculations. In Fig. 31 is shown a drawing of a bent-up detail, and the flat development from which it was produced. As no angles are given

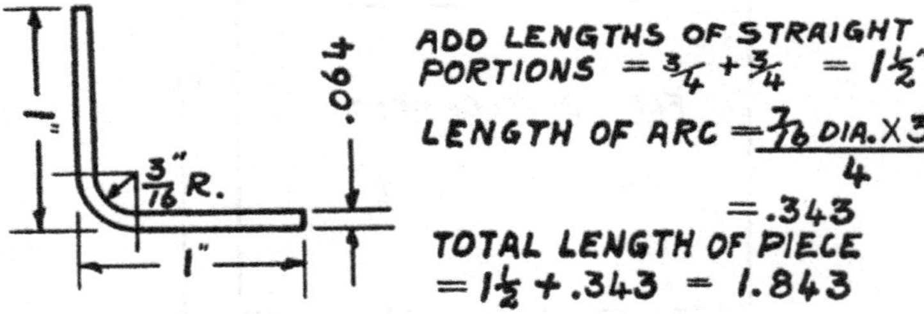

Fig. 32

it is assumed that the bends are 90 deg., or right-angle bends. Fig. 32 illustrates the method employed, but first the apprentice must understand that the calculations are made with reference to the center line of the metal. Obviously, when being bent, the metal on the inside of the radius crowds or shrinks, while that on the outside of the radius stretches. The center line remains neutral.

It will be seen that the over-all dimensions are determined by adding the lengths of the straight portions to the combined length of the arcs. In some cases, however, especially in drawn shells, it is impossible to predict accurately what the developed shape will be, so it is necessary to leave a certain amount for a trimming operation.

Those parts of the airplane that maintain the aerodynamic contours, such as bulkheads and ribs, are made to fit what are known as master templates. These must be made very carefully to represent the contour at the inside of the skin line. The assembly fixtures are then made to fit the master templates. In order to produce a contour on a template, it is necessary to "lay out" the ordinates, as given on the engineering drawing.

Spline—The points thus obtained are connected by a "spline," which is held down by lead weights, then drawn in with a scriber. The spline consists of a long, flexible piece of wood of about 3/16 in. thick, by 3/8 in. wide. A groove

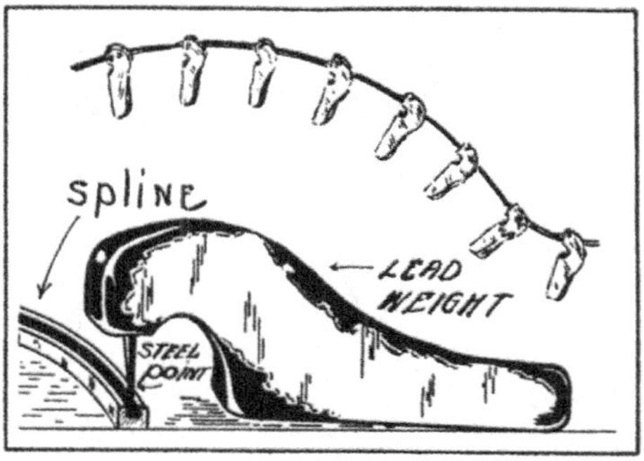

FIG. 33

along its entire length acts as a receptacle for the lead weights (Fig. 33). Very good splines are also made from flexible resinous compounds.

Templates are also made to assist the lathe operator in grinding tools to cut specific shapes or for checking form rolls or form cutters. In making templates for both male and female form rolls, the apprentice must keep in mind the fact that the contour of the working surfaces in both rolls must differ by the thickness of the material to be rolled. This rule also holds good when making templates for checking form punches and dies.

Micrometer—At this point in his career, the apprentice should be able to take micrometer and vernier readings, and understand the use of the sine bar. The micrometer, Fig. 34, is designed to give accurate readings of such things as the diameter of a round bar, the distance between two pins, or the thickness of a piece of sheet metal.

How to Read a Micrometer Caliper—The spindle C is attached to the thimble E, on the inside, at the point H. The part of the spindle which is concealed within the sleeve and thimble is threaded to fit a nut in the frame A. The frame being held stationary, the thimble E is revolved by the thumb and finger, and the spindle C being attached to the thimble revolves with it, and moves through the nut in

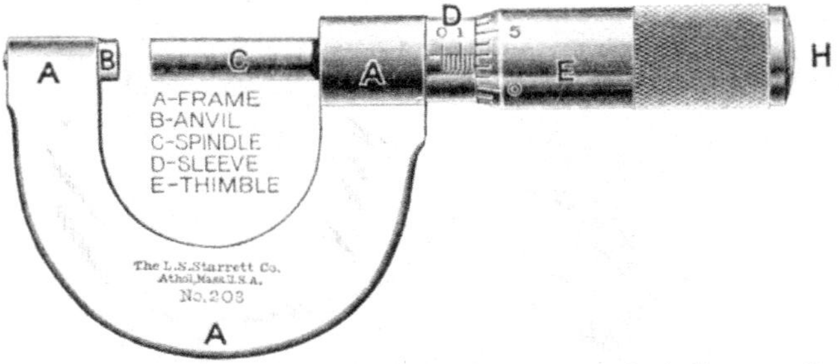

Courtesy L. S. Starrett Co.

FIG. 34—HOW TO READ A MICROMETER CALIPER

the frame, approaching or receding from the anvil B. The article to be measured is placed between the anvil B and the spindle C. The measurement of the opening between the anvil and the spindle is shown by the lines and figures on the sleeve D and the thimble E.

The pitch of the screw threads on the concealed part of the spindle is 40 to an inch. One complete revolution of the spindle therefore moves it longitudinally one-fortieth (or twenty-five thousandths) of an inch. The sleeve D is marked with 40 lines to the inch, corresponding to the number of threads on the spindle. When the caliper is closed, the beveled edge of the thimble coincides with the line marked 0 on the sleeve, and the 0 line on the thimble agrees with the horizontal line on the sleeve. Open the caliper by revolving the thimble one full revolution, or until the 0 line on the thimble again coincides with the horizontal line on the sleeve; the distance between the anvil B and the spindle C is then 1/40 (or .025) of an inch, and the beveled edge of the thimble will coincide with the second vertical line on the sleeve. Each vertical line on the sleeve indicates a distance of 1/40 of an inch. Every fourth line is made longer than the others, and is numbered 0, 1, 2, 3, etc. Each numbered line indicates a distance of four times 1/40 of an inch, or one-tenth.

The beveled edge of the thimble is marked in twenty-five divisions, and every fifth line is numbered from 0 to 25. Rotating the thimble from one of these marks to the next moves the spindle longitudinally 1/25 of twenty-five thousandths or one-thousandth of an inch. Rotating it two divisions indicates two thousandths, etc. Twenty-five divisions will indicate a complete revolution, .025 or 1/40 of an inch.

To read the caliper, therefore, multiply the number of vertical divisions visible on the sleeve by 25, and add the number of divisions on the bevel of the thimble, from 0 to the line which coincides with the horizontal line on the sleeve. For example, as the tool is represented in the engraving, there are seven divisions visible on the sleeve. Multiply this number by 25, and add the number of divisions shown on the bevel of the thimble, 3. The micrometer is open one hundred and seventy eight thousandths ($7 \times 25 = 175 + 3 = 178$).

Vernier—The vernier system of measuring employs an ingenious method by which it is possible to measure in thousandths of an inch from a straight scale. The discoverer of this method was M. Pierre Vernier, a Frenchman.

How to Read Height Gauge or Caliper with Vernier—The bar of the tool (Fig. 35) is graduated in fortieths or .025 of an inch, every fourth division, representing a tenth of an inch, being numbered. On the vernier plate is a space divided into twenty-five parts and numbered 0, 5, 10, 15, 20,

25. The twenty-five divisions on the vernier occupy the same space as twenty-four divisions on the bar.

The difference between the width of one of the twenty-five spaces on the vernier and one of the twenty-four spaces on the bar is therefore $\frac{1}{25}$ of $\frac{1}{40}$ or $\frac{1}{1000}$ of an inch. If the tool is set so that the 0 line on the vernier coincides with the 0 line on the bar, the line to the right of the 0 on the vernier will differ from the line to the right of the 0 on the

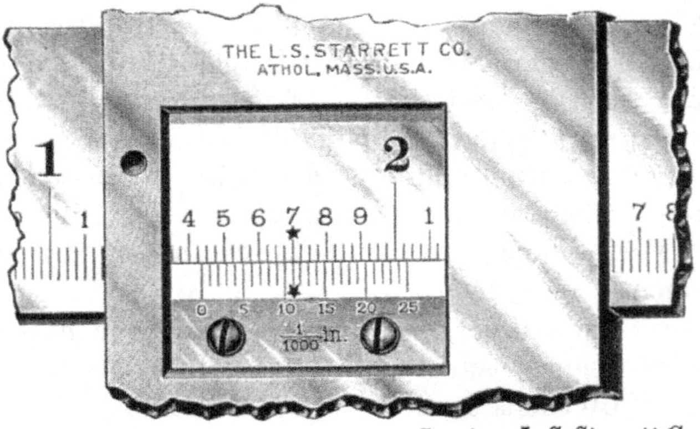

Courtesy L. S. Starrett Co.

FIG. 35—HOW TO READ HEIGHT GAUGE OR CALIPER WITH VERNIER

bar by $\frac{1}{1000}$ of an inch; the second line by $\frac{2}{1000}$ of an inch and so on. The difference will continue to increase $\frac{1}{1000}$ of an inch for each division until the line 25 on the vernier coincides with the line 24 on the bar.

To read the tool, note how many inches, tenths (or .100) and fortieths (or .025) the 0 mark on the vernier is from the 0 mark on the bar; then note the number of divisions on the vernier from 0 to a line which exactly coincides with a line on the bar. *Example:* In the above cut the vernier has been moved to the right one and four-tenths and one-fortieth inches (1.425), as shown on the bar, and the eleventh line on the vernier coincides with a line, as indicated by the stars, on the bar. Eleven thousandths of an inch are therefore to be added to the reading on the bar and the total reading is one and four hundred and thirty-six thousandths inches (1.436).

The most common applications of the vernier system—the height gauge and the vernier calipers—are shown in Fig. 36.

Courtesy L. S. Starrett Co.

Fig. 36

1. Height Gauge
2. Vernier Caliper

Sine Bar—The sine bar is yet another precision tool with which the apprentice must familiarize himself. This tool, Fig. 37, is used for checking angles; therefore, a knowledge of elementary trigonometry is absolutely necessary. The diameter of the two round plugs must be identical, and their center distance, B, should be either five or ten inches, for ease in making trigonometric calculations. As can be seen by the illustration, if the sine bar is clamped to a piece of steel with an angle to be checked, the difference between the two height-gauge readings will give the length of side A,

Side B is known to be five or ten inches, according to the length of the sine bar, so the sine of the angle X is known to be A/B.

Making a Drill Jig—From the templating group, the apprentice may be put to making small drill jigs. Although the function of the drill jig has already been described, a few remarks from the point of view of the tool maker may be helpful. First of all, he must study the tool drawing

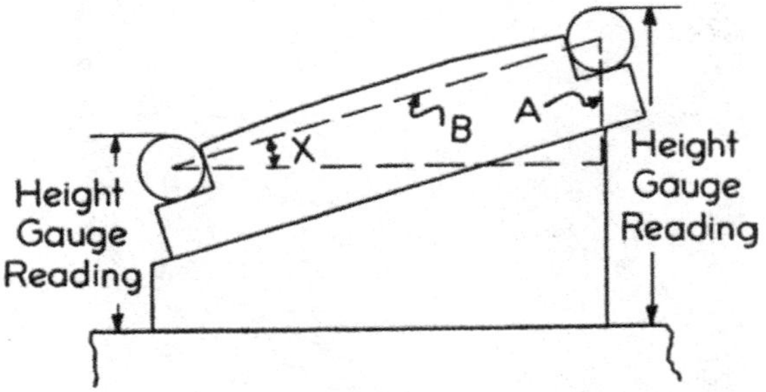

Fig. 37—Application of Sine Bar

to familiarize himself with the design and the function of each detail; then he must take steps to obtain the necessary steel, cut to the proper length. The round material for bushings or pins should be left in the bar lengths, as the cutting off is done by the lathe operator after all the machining is completed. In a properly organized toolroom, the steel will be delivered to the toolmaker, cleaned and burred (rough edges removed); then he will have the ends squared up on the mill or shaper ready for assembly. The structure will then be clamped together so that it may be arc-welded.

Materials—When a welded joint has to be machined over later, it is advisable to bevel the steel at the points of contact, so that the weld will not be removed by the machine operation (Fig. 38). There are two types of steel commonly used in making drill jigs, that which is hot-rolled and that which is cold-rolled. Both have certain advantages. The former carries very little internal stress, due to its extreme ductility during the rolling operation. The bar dimensions, however, are not very accurate, and the surface is rough, but this is offset by the fact that machining produces a minimum of distortion.

Cold-rolled steel, on the other hand, holds its nominal size quite accurately, usually within .002 in., but due to the high pressure to which it is subjected while being rolled, a "skin" is produced on all the rolled surfaces, holding the

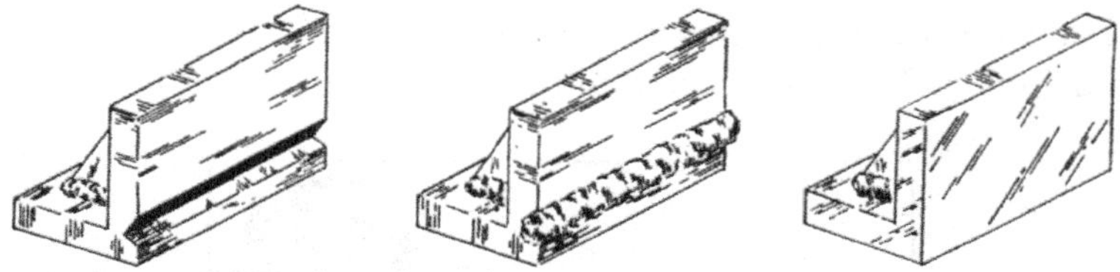

FIG. 38

metal straight and in equilibrium. Any subsequent removal of the "skin" by machining destroys this equilibrium and causes the bar to curve inward towards the machined surface. By normalizing (annealing), the surface strains just described are removed, and the bar may be machined without producing distortion.

How important it is that the toolroom apprentice should learn to use sound judgment in the selection of steel. A great deal of responsibility, however, rests with the welder, because, by localized overheating and by piling the welds too high, he can produce strains that will cause distortion even in hot-rolled steel.

After machining the important surfaces, the jig is ready to have the holes bored to receive the drill bushings. This is accomplished on the jig borer, a description of which will follow. The bushings, which have been hardened and tempered, must now be ground to fit the bored holes. In grinding bushings, it is extremely important that the outside diameter be concentric with the diameter of the hole. If this is not the case, the hole in the finished part will be off location when drilled in the jig.

If a reamed hole is called for, two bushings are necessary, one for the drill and the other for the reamer, and they must be ground to a slip fit so that they may be removed by hand. A knurled head is provided to facilitate their removal. The outside diameter of a slip-fit bushing should not be more than three tenths of a thousandth of an inch undersize from the hole in which it has to be inserted. Lock screws must also be provided (Fig. 39) so that the bushing will not revolve due to the action of the drill.

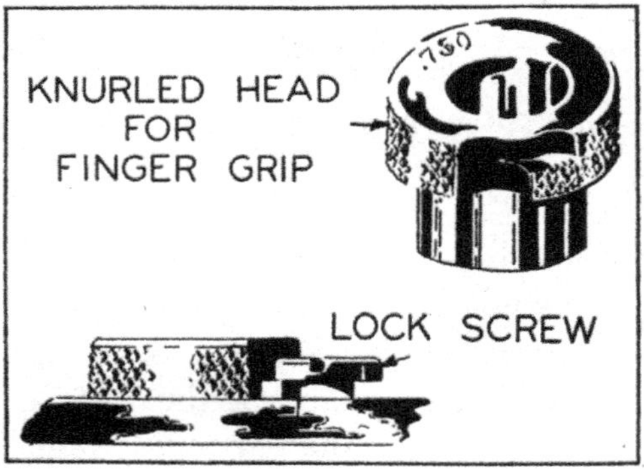

Fig. 39

When high production is required of a drill jig, it is a good practice to provide hardened-steel liners with the slip bushings, so that their frequent removal will not cause undue wear. When the finished piece has to be drilled without reaming, then press-fit bushings are inserted into the jig. These should be ground to approximately five tenths of a thousandth oversize, and pressed into place.

After proper clamping facilities have been provided, the jig is ready to be sent to the tool inspection department for checking. The foregoing description of the manufacture of a drill jig is very general in character. In actual practice it will be found that each jig will present different problems that experience alone can solve. The apprentice must learn to accomplish all the operations just mentioned, including grinding, heat-treating and welding.

Jig Borer—At this time some mention must be made of that amazingly accurate machine tool called the jig borer. It was purposely omitted when the machine-shop tools were described, as it is essentially a toolroom machine.

On this machine it is possible to bore holes the diameters of which must be within tolerances of two tenths of a thousandth of an inch, and to bore a series of holes the center distances of which must be held within five tenths of a thousandth. The spindle of this machine is vertical, a transmission box and lever making possible the different spindle speeds. The work-supporting table is very accurately leveled up to assume a 90-deg. relationship with the spindle. The table can also be moved by handwheels in two directions, the axis of both movements being exactly 90 deg. to one another. In this way, any point on the table can be brought directly under the axis of the spindle. The frame of this machine must be strong and rigid because any misalignment between the table and the spindle would render the machine useless as a precision machine tool.

In order to establish definitely the boring position with micrometer accuracy, two indicator dials are used, one governing each directional movement of the table. A groove between the table and each indicator is filled with a series of end measuring bars of known length, one of which is adjustable to a micrometer reading. As an example, suppose the operator wishes to move his table a distance of 7.325 in. in one direction from its existing location: First of all he notes the reading on the adjustable bar, then he rotates the face of the dial indicator until it reads zero. The table is then moved by means of a handwheel a distance which is slightly greater than 7.325 in., so that additional bars amounting to 7 in. can be inserted in the groove. Then the adjustable bar is increased in length by .325 in. and put back into the groove, after which the table is moved back until the dial reads zero. The movement of the table and also the work which is clamped to it is therefore known to be 7.325 in. It must be mentioned, however, that a continued pressure on the dial plunger will cause the pointer to rotate several times around the dial. In order to avoid mistakes, it is advisable that all settings be such that zero may be reached with a pointer movement of less than one revolution.

In order to control the diameters of the holes being bored, a special chuck allows the boring bar to be moved radially by a micrometer setting, thus giving the operator perfect control of the amount of cut.

Dies—It is perhaps in the manufacture of drawing and forming dies that the tool maker attains his highest degree of craftsmanship. This observation is made because no data are available which set forth *all* the processes involved. A die may conform to its tool drawing and to all the known standards of diemaking, yet not produce a satisfactory part. A few deft touches, however, drawn from the storehouse of experience and intuition, very often get production going. For instance, when difficult shapes are being drawn in two operations, it will sometimes happen that the final draw will produce a fracture which the tool maker must get rid of. He usually does this by contriving in the first operation to gather surplus metal in such locations that it will "flow" readily into the possible fracture points, thus producing a final sample that is free from strain.

Fits—It is also essential that the toolroom apprentice should learn something about fits—not that he should take a first-aid course, for the fits referred to concern the necessary relationship in size between two pieces of metal in order to produce a desired result. In most cases the work in this category is round, the desired fit being the degree of tightness of a round spindle in a hole. The apprentice, after having gained a little experience, will have to prepare his own sketches for machine work, so he must learn how to specify various fits.

Limits and Allowances—First, he must determine the allowance necessary between the two pieces to produce the necessary fit. For instance, the "running fit" must allow for a film of oil between the spindle and the hole and may be from .0005 in. to .0025 in. He may then specify the limits of size for the hole and spindle, so that if the hole is made to the smallest possible size and the spindle to the largest, the allowance will be .0005 in. If the spindle is made to the small limit and the hole to the large, the allowance will be .0025 in. If a "push fit" is desired, the allowance will not be so large, so that the specified limits for the spindle and hole will have to be specified in tenths of a thousandth. Such accuracy requires the use of an external and internal grinder, which, of course, are essential parts of the toolroom equipment.

The "drive fit," on the other hand, requires that the spindle be larger than the hole and that it cannot go together by hand. An arbor press should be used and the surfaces of the spindle and hole should be oiled so that the pores of each will retain enough lubrication to prevent their "freezing" together.

When the "force fit" is used, the difference between the diameters of the spindle and the hole are so great that heat must be used to expand the hole before the spindle can be driven home. With cooling, the tremendous power of the contraction renders it very difficult to separate the two pieces.

All the necessary inspection and checking gauges are made in the toolroom, requiring a high degree of skill and accuracy. But whether the tool maker is setting two locating points many feet apart in a large assembly fixture, weighing several tons, or making a small plug gauge weighing only a few ounces, he must work with a precision that comes from long practice and from his own inherent desire to work that way.

Toolroom Machines—The toolroom machines are similar to those of the machine shop, but certain necessary additional machines are listed below.

Jig borer
Contour saw and filing machine (equivalent to woodworking band saw)
Bench lathe
Duplicator (reproduces irregular shapes in steel from models)

CHAPTER 9

PRINCIPLES OF SHEET-METAL WORK

The manipulation of sheet metal by hand is one of the oldest of crafts. In satisfying the inherent human desire for ornamentation, the sheet-metal worker of centuries ago developed a manual technique that still lingers. As industry developed and the necessity for speed in manufacture grew, the hand methods were gradually replaced by machine operations.

In some industries, the sheet-metal craftsman has been eliminated entirely, the machines producing in such quantities and in such beauty of form and symmetry that sheet-metal objects have become an essential part of our civilization. Such things as automobiles, refrigerators, office furniture and cabinets of all kinds are proof of the wide variety of the modern application of sheet metal.

In the aircraft industry, sheet metal is used very extensively. In fact, it may be said that the completed airplane, apart from the engine, is 90 per cent composed of fabricated sheet-metal parts. Machines have been utilized to a very great extent in accomplishing this, but there comes a point where the machine fails, and the sheet-metal worker steps into the picture. Like the craftsman of old, he beats the metal to do his bidding, with a skill that was born when the flight of man was still just a dream of the ancient poets.

Sheet-metal Apprentice's Tool Kit—The sheet-metal apprentice who wants to make a good start should report for work with the following tools:

1. Ball-peen hammer (8 oz.)
2. Planishing hammer

3. 6-in. flexible scale
4. 12-in. combination square with protractor
5. Center punch
6. Tinsnips
7. Mallet (wood, fiber or rubber)
8. Dividers
9. Small C clamps (2 pair)
10. Flexible six-foot rule
11. Pliers

Let us consider first of all that phase of aircraft sheet-metal work that is accomplished by machine. In order to do this, however, it will be necessary to acquaint the apprentice with the various metals that are used in aircraft work.

The characteristics of these metals in most cases determine the method of forming and sequence of operations.

Aluminum and Its Alloys—Aluminum, the principal basic metal used in aircraft construction, is very easily manipulated, but it must be alloyed with various other metals to give it the necessary strength. Aluminum and its alloys are known in this country by certain code designations which have been specified by the Aluminum Company of America. The code is divided into three parts, as for example 24ST. The number signifies what alloying materials have been used. The letter S signifies that the resulting metal is a wrought aluminum alloy, and the letter T signifies that it has been heat-treated. The example 24ST is a wrought aluminum, magnesium, copper, and manganese alloy in its heat-treated state. The other heat-treat designations are listed below.

O—Annealed.
H—Hard-rolled or hard-drawn from the annealed state.
T—Maximum heat-treat.
W—Heat-treated but not aged.
RT—Heat-treated, aged and then cold-worked.
F—As fabricated.

Some examples of the complete codes are given below.

2SO—No. 2 alloy, annealed.
2S½H—No. 2 alloy which has been cold-worked after annealing, enough to raise its temper to half the full extent.
24SO—No. 24 alloy, annealed.
24ST—No. 24 alloy, heat-treated and aged.
24SRT—No. 24 alloy, heat-treated and cold-worked (as by rolling).
52SF—No. 52 alloy, as fabricated.
53SO—No. 53 alloy annealed.
53SW—No. 53 alloy, heat-treated but not aged.
53ST—No. 53 alloy, heat-treated and aged.

In order to render these symbols more understandable, some further explanation is necessary. The wrought aluminum alloys are divided into two classes, those which may be hardened by heat-treatment and those which may not. The purpose of the heat-treatment is to increase the tensile and shearing strength of the material, and consists of heating to a specified temperature in a bath of sodium nitrate. This temperature is held for a certain length of time, after which the material is quenched as rapidly as possible in cold water.

In the case of 17S, the specified temperature is from 930° F. to 950° F., and the length of time at this temperature is from 5 to 30 minutes, depending on the thickness of the material. For 24S, the temperature should be brought to 910° to 930° F., and for material up to ⅛ in. thick, the soaking time is 15 minutes; over ⅛ in. and up to 3/16 in. it is 20 minutes; over 3/16 in. and up to ½ in. thick, it should be soaked for 30 minutes.

After quenching in running cold water, the material is in the W condition, which means that it is ready to start aging. Not until the aging has been completed does the material attain its full tensile strength.

Aging—The aging process differs somewhat with the different alloys; for instance, 17S and 24S require no further treatment, but start to age-harden spontaneously at room temperature. Within 24 hours they attain 90 per cent of their full strength, and within four days they reach their maximum tensile and shear characteristics. In this state they are known as 17ST and 24ST.

The alloys known as 53S and 61S, however, require artificial aging. They must be heated to approximately 320° F., and held at this temperature for 18 hours, before attaining the 53ST and 61ST condition.

Those alloys which do not respond to hardening by heat-treatment, such as 2S, 3S, 4S, and 52S, attain their maximum temper by the cold working they receive during manufacture. This may be either rolling or drawing. The degree of temper is designated by the symbol H, ¼H, ½H, or ¾H. Such alloys, however, *do* respond to the annealing process.

It may be noticed that some codes are preceded by the letters ALC which signify that the alloy is covered on both sides with a coating of pure aluminum, and is known as Alclad. This coating prevents corrosion and amounts to approximately 10 per cent of the thickness of the sheet.

The alloy already mentioned as 24S possesses very desirable characteristics (tensile strength of 62,000 pounds per square inch in the T condition, and 65,000 pounds per square inch in the RT condition), so it is used very extensively in the aircraft industry. After heat-treatment it also attains a very high corrosion-resistance value.

In forming this metal, it is important for the apprentice to note that during the first few hours of the aging process (W condition) it is extremely soft and ductile. In fact, during the first 20 minutes of aging, the material is softer than when in 24SO. Any forming that is so severe as to produce cracks if done in 24ST should therefore be done with SO material, and any buckles or distortion due to the heat-treatment may be removed during the first 20 minutes of the W condition.

Many straight bends and even bends with offsets may be successfully accomplished, however, in 24ST, if sufficient bending radius is used and if the excessive spring back is allowed for in the tools.

Drop Hammer—The machine that accomplishes most of the difficult forming jobs is the drop hammer, which consists of a solid cast iron bed for supporting the dies and the work and two upright slides or ways, between which the huge hammer drops to deliver its blow or rises again to the clear position. The hammer itself may be actuated by an air cylinder or merely raised and released by a few coils of rope around a constantly revolving drum. A slight pull on the rope by hand is enough to generate sufficient friction around the drum to lift the enormous weight of the hammer. The relaxation of the rope causes the hammer to drop by its own weight. The drop-hammer die is made from zinc, having enough body to resist the terrific pounding it receives, and being molded to shape from a plaster model. The punch which engages the die is made from a lead alloy, and, having a lower melting temperature than zinc, it is poured as a liquid into the die to assume its shape when cooled. An amount equal to the thickness of the metal to be formed is scraped off the punch while it is being dressed. Most of the work done by the drop hammer is of such severe contour that it is best done in 24SO, then heat-treated and restruck during the early stages of the W condition. By placing rubber pads on the die at strategic points for the initial strike, the restriking operation may be greatly facilitated. Where this is not possible, it is sometimes necessary to provide for finishing off by hand at the bench or on the power hammer.

In all cases a checking block is necessary so that the accuracy of the piece may be determined. This block may be composed of reinforced plaster or wood.

Plaster Molds—A trip to the department where the plaster molds are made will indicate to the apprentice that here indeed are the "sculptors of modern industry." The graceful flowing contours of the airplane are accurately reproduced in sweeping panels of snow-white plaster. Left- and right-hand curves are produced with amazing fidelity to blueprint specification.

The pieces thus formed in the drop hammer must, of course, be trimmed to the proper outline. For this opera-

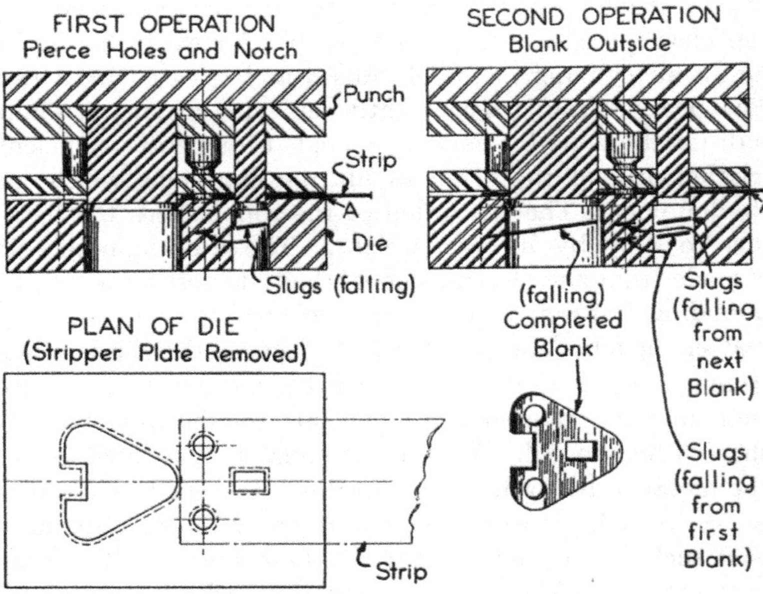

FIG. 40

tion, trim templates are made in the same dies, the trimming being done by band saw and the rough edges smoothed by a special hand tool.

Punch Presses—In the punch-press department may also be seen some very interesting aspects of sheet-metal work. Long rows of varying sizes of presses are lined up, from the small five-ton, quick-action machine to the 3000-ton, hydraulic monster.

Here is a very good example of the interconnection of one department with another. The dies used in these machines are designed by the tool engineer and made by the tool maker, but as they are set up and operated in the sheet-metal department, it is deemed advisable to describe something about them in this chapter.

Blanking—In front of each small machine is seated a man who seems to be feeding an insatiable maw with steel. What is he doing? You have to come close to see. The ram of the nearest press reciprocates vertically with short staccato strokes. It is difficult to see just what takes place when the thing comes down and clicks, but you see that a

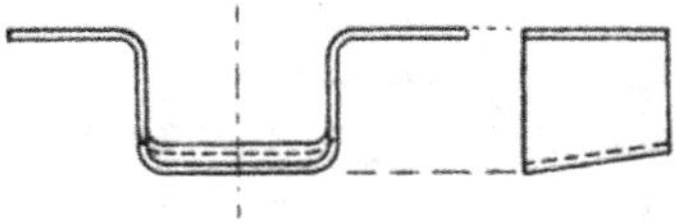

FIG. 41

piece of steel drops into a pan with every click. Pick one of them up, and study it. Yes, a nice little job, and to think that with one stroke of the press, this piece was blanked to shape and two holes and a square notch put in. Actually two strokes of the press are required to produce this blank (Fig. 40), although one drops out at each stroke. The die that accomplishes this is called a progressive blank and pierce die, and is shown in Fig. 40.

The strip steel is entered at A, and at each stroke of the press, as can be seen by the sketch, the holes are pierced in the strip, and the preceding piece is blanked out. By this method, two operations are completed at one stroke of the press, resulting in one finished blank with each click.

Forming—The next press is doing a forming job, which means that a flat piece of steel is being bent to the required shape (Fig. 41). A drawing of the form die is shown in Fig. 42, in which A is the punch, B is the die, and C, the pressure

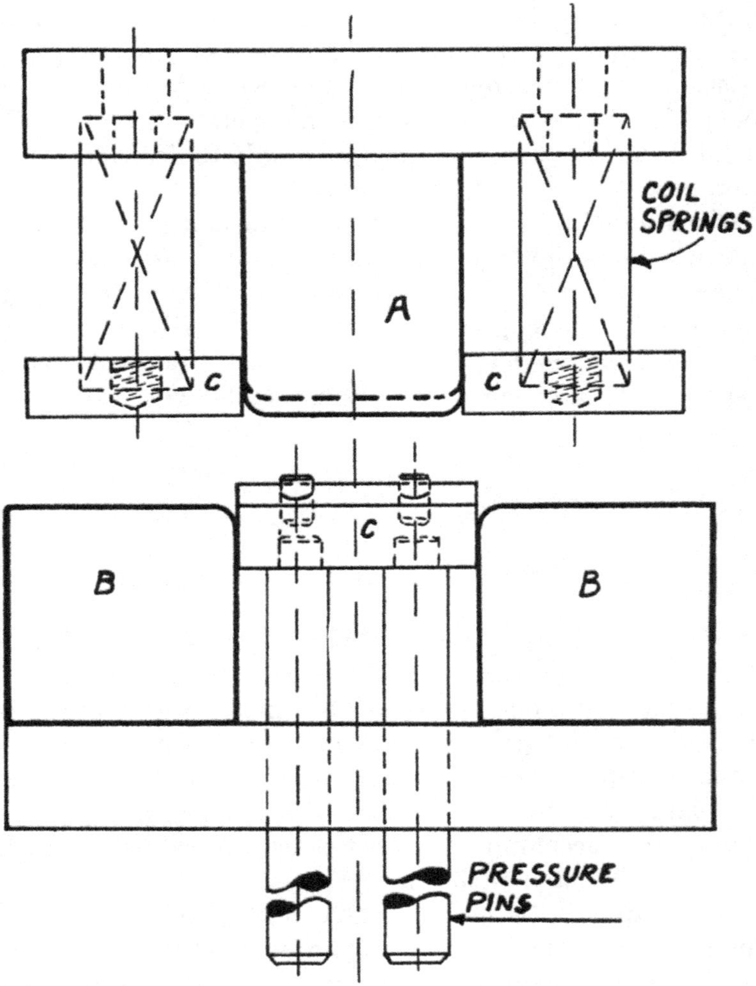

FIG. 42

pads. The piece of steel to be formed is placed on the lower pressure pad and is located by means of two pins which fit into the two locating holes in the blank. When the punch comes down, it pushes the blank and the pad towards the bottom of the die, which results in the blank's taking the exact shape of the punch. The pressure is applied to the lower pad by means of a column of rubber, which exerts an upward pressure on a steel plate upon which rests the pressure pad pins.

The more modern trend is to utilize air pressure in place of the rubber, the air cylinders being located below the bolster plate of the press.

Pressure Pad—In the proper application of punches and dies, the pressure pad plays a very important part. In the die we have just described, what would be the result if no pressure pad were provided? Fig. 43 answers this question. The punch descends and comes in contact with the blank. It continues downward and causes the blank to

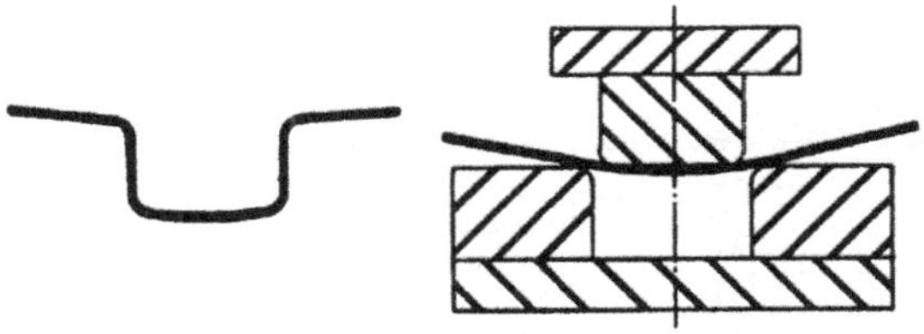

FIG. 43

spring away from the punch as shown. This destroys the 90-deg. relationship between the sides of the piece and the base. Although the piece is brought instantaneously to its proper form when the punch is at the bottom of its stroke, the spring back that was initiated at the beginning of the operation will result in the undesirable finished form as shown in the sketch.

Draw Dies—In the punch-press department you will also find various shapes of dished forms being produced from flat blanks. The die that accomplishes this is called a draw die. In designing such a die, some very important factors and their relationship to one another must be considered. There is the ratio between the depth of the draw and the diameter of the required dish. Naturally, the deeper the draw, the more difficult the operation. Then there is the ductility and inherent strength of the material to be drawn. For instance, shallow forms can be completed in one operation, but deep shells or cylinders require two or three operations, depending on the factors just mentioned. Too much stretching of the metal will cause it to tear.

A description of the action of the die shown in Fig. 44 may help to clarify these statements. The round blank is placed on top of the pressure ring as shown in the open position. The punch or forming ring descends, gripping the

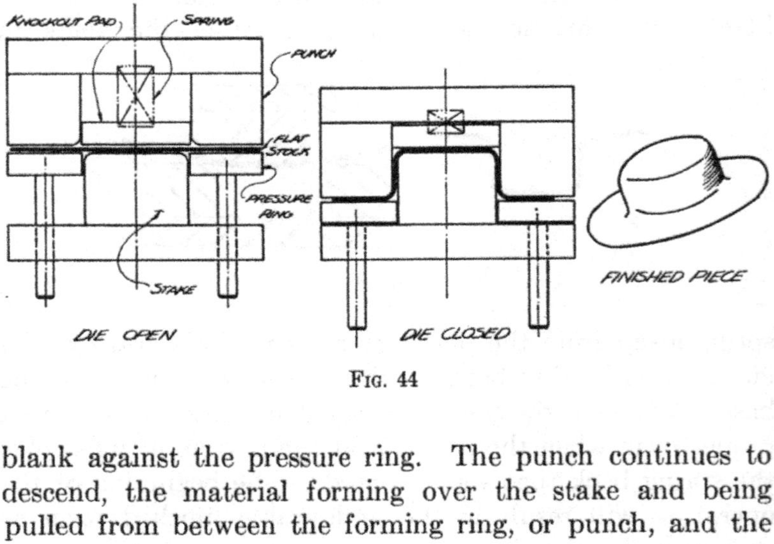

FIG. 44

blank against the pressure ring. The punch continues to descend, the material forming over the stake and being pulled from between the forming ring, or punch, and the pressure ring.

During this operation, the material has a tendency to wrinkle, because of the small diameter of the stake receiving the larger diameter of the blank around it. The pressure between the two rings has the effect of ironing out these wrinkles, but care must be taken to insure that this pressure will not produce too much drag on the material, thereby causing it to fracture as it forms over the radius of the stake. The knockout pad shown in the sketch is for ejecting the formed pieces, as they will have a tendency to stick inside the forming ring.

When two operations are used, the shell is first drawn to a larger diameter but more shallow than the finished piece. The redrawing operation finishes the piece to the required dimensions. The use of a thin coat of lubricant over the surface of the blank will reduce the pressure drag to a minimum. When aluminum alloy is being drawn, a lard-oil lubricant is very satisfactory. The radii on the die must be highly polished in the direction of the draw, otherwise the die becomes "loaded" or impregnated with particles of aluminum which scratch the surfaces of the pieces being formed. In this connection it must be remembered by the man in the shop that when a piece of aluminum or any of its alloys become scratched, it is rendered absolutely useless. If such a piece were installed in an airplane the vibration and strain would cause the scratch to increase in depth so much that it would develop into a fracture. A steel scriber, therefore, must never be used to draw lines on these metals. It is safer always to use an ordinary lead pencil.

In making deep-drawn parts, it is advisable to use material which is about .002 in. or .003 in. thicker than that called for on the engineering specification. This allows for thinning of the material due to the extreme tension to which it is subjected while flowing over the forming radii. In setting up these dies, the operator requires a great amount of skill in maintaining perfect alignment and in using the correct amount of ram pressure.

Rubber Dies—The large hydraulic presses are for use with the rubber die, which consists of a large rectangular slab of live rubber about three inches thick, confined on all four sides by a heavy steel container. Being mounted on the ram of the press, this rubber slab descends upon a solid steel plate. After coming in contact with the steel plate, the pressure on the rubber begins to mount very rapidly, until it reaches a predetermined figure at which it is shut off automatically. The ram of the press then ascends. If a steel form block, with a suitably located blank on top of it, is placed on the solid steel plate, the descending rubber will form the blank to shape, just as if a steel die had been used for the operation. Blanking may also be done in this manner, if a cutting template with sharp edges is used.

Courtesy Douglas Aircraft Corp.

FIG. 45A—SHOWING WORK BEFORE RUBBER DIE DESCENDS

1. Ready for blanking
2. Ready for forming

The economic use of the rubber die is clearly shown in the photographs (Figs. 45 A and B), in which 16 blanks are ready to be formed or blanked. The result of this same operation is clearly seen in Fig. 45 B, after the rubber has done its work. It is possible to form pieces in this manner so accurately that no trimming operations are necessary.

Power Brake—The power brake is another machine that figures very largely in sheet-metal operations. In this machine, flanges may be bent to any degree for a length of 6, 8, or even 14 feet, depending on the capacity of the machine. Almost all bends that are free from contour can be made in the brake, and with the use of special dies it is possible to produce a corrugated sheet. This is done by forming one corrugation with each stroke of the machine. While the flat sheet enters the die, it emerges as a corrugated sheet. The smaller brakes, up to 4 feet long, are operated by hand, but their application is limited to the folding up of flanges.

Courtesy Douglas Aircraft Co.

FIG. 45B—SHOWING WORK AFTER DIE DESCENDS

1. After blanking (surplus metal must be removed)
2. After forming

Stretching Machine—Yet another machine, the stretcher, has been recently developed for the aircraft industry, and its purpose is to form large panels of skin, or engine cowling, to the proper shape. A flat, rectangular sheet of 24ST material is firmly clamped along two opposite edges, and a hydraulically operated ram, bearing a large form block, is raised up so as to come in contact with the sheet. A continuation of the upward movement of the form block, with the sheet firmly held down along two edges, causes the sheet to be stretched over the form block and to take the same shape as the block. After this it must be trimmed.

Bench Work—The sheet-metal bench hand depends more than any other aircraft worker on his manual skill. It is he who adds the finishing touches to those parts that cannot be formed adequately by machine. If any parts are required in such small quantities that special tools are unwarranted, he is expected to produce them by hand. This is very often the case in experimental work. It is therefore essential that he understand geometrical layout thoroughly, for he must develop in the flat many peculiar shapes. When making his layout on the metal, he must adhere to the rule already mentioned of never using a steel scriber, so as not to scratch the material. His principal problems are in shrinking or stretching the metal. The apprentice must remember that if sheet metal is bent and in so doing must occupy a smaller area than it originally had, it must be shrunk, or it will develop wrinkles. If it must occupy a larger area, it must be stretched, or it will crack and split open. The sheet-metal worker knows that when shrinking metal, he must either gather or deliberately crimp the surplus caused by the actual forming and proceed to hammer it flat or to any desired shape.

While doing this it may be necessary for him to restrict the effect of his hammering by rigidly clamping each side of the area on which he is working.

In stretching metal he must be careful not to hammer so much that the metal becomes too thin and too brittle. Every hammer blow must count and must be aimed at the correct spot. It is also important that every blow should be properly backed up by a form block or by some special "dolly" from the personal tool kit of the sheet-metal worker.

The sheet-metal apprentice will also find out that he must learn how to cut sheet metal with a pair of tinsnips. He may have been able to cut paper dolls in his more youthful years, but he will find that in cutting metal the scrap has a habit of getting in the way and sometimes cutting the hands. He must learn that sheared metal cuts readily, and cuts deep. The solution to this problem is in using the proper type of tinsnips for the job and in getting some practice in using them.

As may be gathered from this chapter, sheet-metal work in aircraft is divided into two classes, machine work and bench work. It must be truthfully stated that as time progresses the scope of the machine increases, which continually reduces the necessity for hand operations.

Sheet-metal Storage—The storage of sheet metal is a very important matter in the aircraft factory. There are so many different alloys, with varying heat-treatments, that their withdrawal from the stores must be watched very closely. The wrong material in an important structural member may have disastrous results.

It is also a good plan to protect sheet stock before storing by spraying it with a protective lacquer or by placing paper between the metal sheets. This reduces the scrap loss due to scratching while in storage or in transportation from one department to another.

Sheet-metal Machines—In the sheet-metal department, the machines are classified in three groups: metal forming, metal cutting, and hand machines.

Metal forming
Punch press (mechanical)
Punch press (hydraulic)
Drop hammer
Power brake
Shrinking machine
Swaging machine
Stretching machine
Forming rolls
Beading machine
High-speed hammer

Metal cutting
Power shear
Band saw
Router
Shaper (similar to wood shaper)

Hand machines
Electric drill
Grinder
Buffer

CHAPTER 10

BOLTING, RIVETING AND WELDING

In building any mechanical structure, there are three principal methods of attaching metal parts to one another. These are bolting, riveting and welding. Each of these methods possesses good qualities, and in cases where one of them should be used, either of the others would be distinctly undesirable. Bolts have a decided value in that they have a very high shear strength because they can be made from heat-treated steel. To have rivets of this material would be impracticable, owing to the necessity of forming the end of the rivet to a head after it has been inserted in the hole. The bolt may also be unscrewed and the structure taken apart any number of times without much trouble to the mechanic or damage to the hole.

BOLTING

Drilling for Aircraft Bolts—Among his duties in the aircraft factory the apprentice will be called on to drill bolt holes with the use of a hand electric drill. While doing this he must exercise great control over both himself and the machine. The desired result is a hole of such a diameter that the bolt will have to be tapped lightly into place, and one that runs at 90 deg. to the surface of the metal. This is mentioned as the general run of aircraft work. If any other condition is necessary, such as a drive fit or an angular hole, this will be called for specifically on the blueprint.

In order to do this, the apprentice must grasp the electric drill firmly in a manner prescribed by the design of the handle. His trigger finger must be always ready to stop the machine at the first indication of trouble. He must make sure that the drill is at right angles to the job by sighting it from two directions. After some practice the apprentice should develop the sense of maintaining his drill at the proper angle during the entire operation without having to check. He must remember that it is just as important to have the hole emerge at the proper location as it is to start it at the right spot. The drill itself must be ground properly, but this has already been discussed.

Use of Wrenches—After the bolt has been inserted, it is very important that the apprentice should use a suitable wrench, one that will not damage the corners of the nut. A socket wrench is the best type and should be used wherever possible. In inaccessible places an open-end wrench may have to be used, but one should be selected that is a good fit across the flats of the nut. Next comes the problem of tightening the nut. To the skilled mechanic this is no problem at all, but the apprentice is very likely to use too much muscle power and so strain or strip the thread of the bolt. Should this happen the apprentice is cautioned to be sure to remove this bolt and nut and insert new ones. He must learn to use a wrench that does not give too much leverage, and also to restrain his own strength when tightening a nut. There are numerous locking devices to prevent nuts from loosening due to vibration, and it is important that these not be omitted. The most widely used of these are the split cotter, the spring lock washer, and the stop nut.

RIVETING

Rivets are used very extensively in the construction of the airplane, where strong permanent joints are required. They are available in many of the aluminum alloys already mentioned, and great care must be exercised in their heat-treatment, identification and distribution. If these actions are not rigidly controlled, there will be no assurance that the proper rivets have been used for each job, because they all look alike. A rivet is a plain, round piece of metal, one end of which has been upset by impact to form a head of a specified shape (Fig. 46). In use, the rivet is inserted in a hole that is common to the two or more pieces being joined. The head of the rivet must be tight against one of the surfaces; the other end must protrude to such an extent that it may be upset, thus forming another head of specified shape and dimension. The pieces are then riveted together, the load which tends to pull them apart being overcome by the shear strength of the rivet. This type

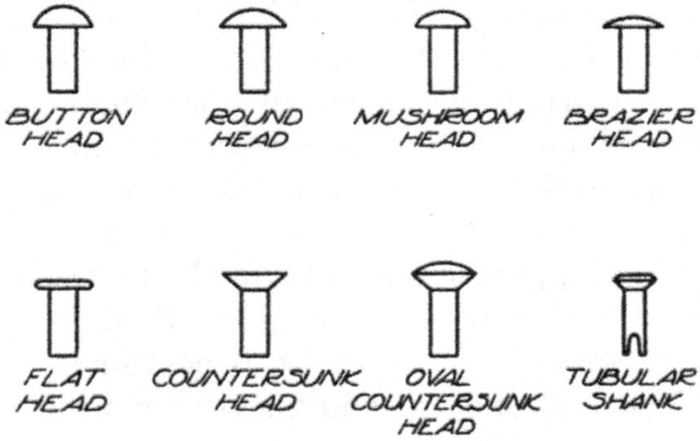

Fig. 46

of joint does not stand up well under a tensional strain, as it is comparatively easy to start a fracture where the underside of the head of the rivet meets the shank.

All aluminum-alloy rivets are cold-driven. When riveting a long row of rivets in a thin sheet there is a tendency for each hole to be expanded by the rivet. This accumulates as each rivet is driven, causing the sheet to buckle. In order to overcome this, a softer modification of 17S, known as A.17S, is used in the manufacture of such rivets. These may be driven after heat-treatment and aging. In the case of larger rivets, where no sacrifice can be made in shear strength, the alloy 24S is generally used. Rivets of this material must be driven while in the early stages of the W condition. It has been found, however, that if held at a freezing temperature after quenching, the W condition is prolonged for at least 24 hours. Full advantage is taken of this additional time by storing the treated rivets in boxes of dry ice at different points throughout the factory. In this way rivets can be treated in greater quantities.

In order to assist the mechanic in selecting his rivets, certain identification marks have been adopted: 17ST rivets are recognized by a slight pimple sticking up from the center of the head, and A17ST by a slight depression. On the head of a 24ST rivet are two small fins, placed radially opposite one another.

If it is necessary to lay out the holes for a row of rivets, a lead pencil should be used, and when center punching, the sheets must be adequately supported to prevent dimpling of the sheet by the center punch. Rivet holes should be only .002 or .003 of an inch larger than the diameter of the rivet. If too much clearance is allowed, the rivet will get out of line with the axis of the driving tool, with the result that the driven head will not be concentric with the shank. With excessive clearance in the hole the rivet may even bend before it is completely headed, resulting in a very weak joint.

When drilling for a rivet with an electric drill, the apprentice must be careful that when the drill breaks through he does not force the rotating chuck of the machine against the sheet he is drilling. His failure to watch this may well scrap the whole sheet.

After drilling for rivet holes, one of the parts must be removed so as to clean the chips away. It is then replaced by inserting screws or skin clamps in a number of the holes. The skin clamp is a clever little device, which, when inserted in a rivet hole, will hold the pieces together in perfect alignment prior to riveting. The proper rivets are then selected and driven home, the clamps or screws being removed one by one.

There are numerous ways of driving rivets, depending on

their type and location. In all cases it is advisable to drive the rivet with as few impact blows as possible. Too many blows subject the rivet to a violent cold-working process which hardens it to such an extent that the head will fracture. The best way to drive an aluminum-alloy rivet is by means of a squeezer. In this machine, a plunger which is actuated by air pressure comes down on the shank end of the rivet. At the same time the yoke of the machine reaches the head of the rivet, thus creating the resistance between the shank and the plunger which causes the head to be formed. Without this resistance, the rivet would be merely pushed out of the hole. This is called "bucking" the rivet. This method is very desirable due to the absence of hammer blows, but unfortunately its application is limited to those rivets that are so near to the edge of the job that they can be reached by the yoke of the machine.

When riveting the skin on the fuselage of an airplane, the men must work in pairs. One mechanic on the inside of the fuselage holds the bucking bar against the rivet, while the other on the outside operates the pneumatic hammer. The latter may be either a "one-shot" impact hammer or a reciprocating hammer, depending on the diameter of the rivets. The power may be applied to either end of the rivet, but in most cases the air hammer is fitted with a rivet peen that fits the original head. When the hammer reciprocates, the shank end of the rivet is driven against the bucking bar. This upsets the end of the rivet and forms head number two.

The importance of the riveted joints of an airplane may well be judged by the fact that each rivet must pass a test by an inspector. He checks the original head to see if it is tight on the surface, and the driven head for diameter and height. Generally speaking, if the shank of a rivet protrudes one and one half times its diameter, this will provide enough metal to form a head that will pass inspection.

The bucking bar must be free from sharp corners or its contact with the sheet will cause damage. The riveter and the one who is bucking should have a pre-arranged signal system if they do not have visual contact with one another. By tapping the structure lightly, the one who is "holding on" notifies the riveter when his bar is in position.

Flush Rivets—In the modern all-metal airplane, the skin rivets should be flush on the outside so as to reduce the parasitic drag by presenting a smooth, unbroken surface to the air stream. This is accomplished by countersinking the relatively thick undermember and "dimpling" the skin sheet to fit the countersink. Great care is necessary in dimpling, so that no cracks occur around the edges of the hole. The dimpling tools must be made very accurately, to take care of the spring back (the skin sheet being tempered) and to reduce to a minimum any distortion of the sheet around each dimple. The rivets used are of the countersunk-head type, and the peen which is inserted in the pneumatic hammer must be flat. In most cases the impact is applied to the countersunk head, and the bucking bar to the shank end, although good results may be obtained by a reversal of this procedure. However, the outside of the airplane is easier to reach with the air hose. In flush-type riveting, very little shearing strain is imposed on the rivet, this being borne by the dimple which is nested in the countersink.

Explosive Rivets—Another very interesting example of riveting practice is the use of the explosive rivet. In confined places which are impossible to reach with a bucking bar, this type of rivet may be used. An explosive charge is carried within the shank end of the rivet so that when a hot iron is placed on the original head, the charge rapidly expands causing the shank end to flare out. This rivet, however, may not be used where great stress is applied. The apprentice must be careful how he handles explosive rivets, as they are a source of danger if left lying around, especially if there is a source of heat in the vicinity.

WELDING

Aircraft metals can also be joined by welding. This method employs heat to melt the contact areas of the metals to be joined, and also the welding rod, which builds up a powerful connection between the two metals. This is known as the fusion method. There are two distinct types of fusion welding, gas welding and arc welding. Although they are the same in general principle, the source of heat is different, and a different technique must be employed.

Gas Welding—Gas welding is particularly satisfactory in joining aluminum and its alloys, so we shall first discuss this method. The gases used may either be oxygen and acetylene, or oxygen and hydrogen. It is usual to store these gases at high pressure in steel cylinders, from which they are conducted to the torch by strong rubber hose. Before entering the rubber hose, however, the gas must pass through a regulating valve so as to reduce the gas from the high pressure in the cylinder to the low pressure necessary at the tip of the torch. The oxygen, hydrogen, and acetylene are stored in separate cylinders, their combination taking place in the mixing chamber of the torch.

It is important that the apprentice should be aware of the danger that surrounds these cylinders if tampered with by an inexperienced person. Oil must be kept away from the regulating valves. If the oxygen at 2000 pounds per square inch comes in contact with the oil, a violent flame is the result, which soon melts the metal of the regulating valve itself.

The welder must learn how to control properly the flame at the tip of his torch, in order to produce a perfect union of the fused metals. The oxyacetylene flame should be adjusted so that it burns with a small white cone about $\frac{1}{4}$ in. long at the orifice of the torch, with a long enveloping flame that is nearly colorless. The oxyhydrogen flame, to produce a good weld, should emerge with a light blue flame about

1 in. long, with an enveloping orange flame. This type of flame prevents the excessive formation of oxide while welding. The torch has two thumbscrews, one for adjusting the

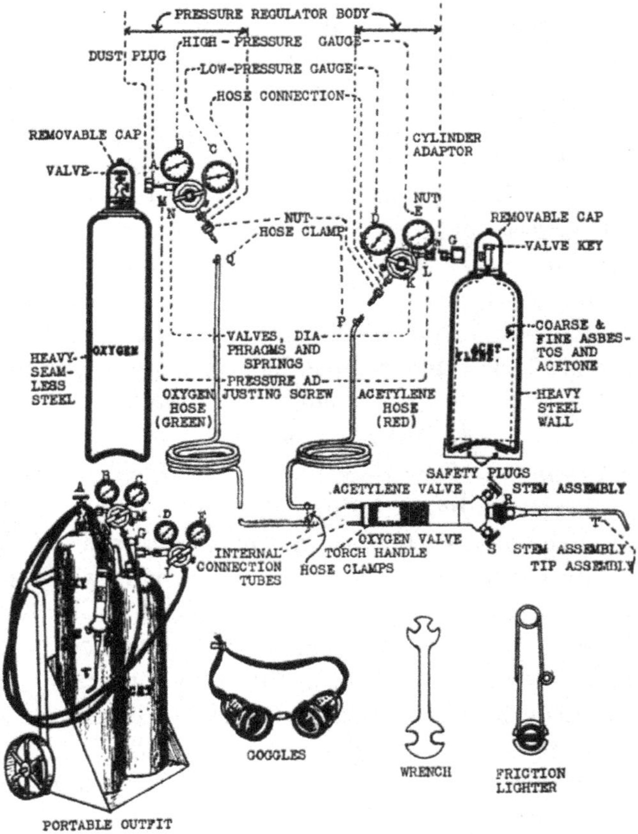

FIG. 47—TORCH-WELDING EQUIPMENT

amount of oxygen and the other for the acetylene or hydrogen.

In order to clear away the natural coating of oxide from the metal being welded so that a sound weld may be the result, a solution or flux, as it is called, must be used. This is in the form of a white paste. If the welding rod is dipped into the paste and then applied to the seam, the heat of the flame will melt the paste and it will run down the rod onto the work. This clears away the oxide film, leaving the metal in a good condition to be fused.

For the gas welding of steel, acetylene is used almost exclusively, as it produces a hotter flame than does hydrogen. No flux is necessary. The tip of the white cone of flame should be brought to within 1/8 in. of the steel to be welded. When the metal reaches melting point the welding rod should be brought into play. As the rod deposits each bead of molten metal, the heat and pressure of the flame push it into place with a weaving, progressive motion. It is only with practice that a proper co-ordination of torch and welding rod can be attained. For these welding operations, dark goggles must be worn, not only for protection of the eyes but to enable the welder to see what is happening at the end of his welding rod.

Arc Welding—The arc-welding process, as the name implies, utilizes the heat of an electric arc to melt the metal. Its use in the aircraft factory, however, is at present limited to the fabrication of steel jigs, for which it is well adapted. The localization of the heat and the speed with which the metals melt under its influence contribute much to reduce distortion. The welding rod in this case is termed the electrode and is coated with flux to within 1½ in. from the end. This uncoated end is held by the electrode holder which has a spring clamp action. Gripping the electrode holder, the welder strikes the arc by drawing the end of the electrode for a short distance along the seam to be welded. When the arc flashes, he continues to weave the molten metal to a solid, penetrating weld. If the electrode is held too close to the work, they will adhere, thus extinguishing the arc. The length of the arc should not be more than 3/16 in., otherwise it will be too unsteady to permit proper control. So excessively brilliant is the arc that the entire face must be covered with a mask, and a glass window through which the welder sees is so darkened that bright daylight is entirely obliterated.

Spot Welding—The spot welding of aluminum alloys has of recent years been developed to a high degree of efficiency in the aircraft industry. During this process, the source of heat is not visible as with a flame or electric arc. Actually, when two pieces of aluminum alloy are being spot welded they are compressed between two electrodes or tips and, by the momentary passage of an electrical current of high intensity, the sheets are fused together. It is interesting to note that the melting begins where the two sheets meet and not, as would be supposed, where the electrodes touch them. The electrodes are made from special alloy copper and are water cooled. So as to produce good welds, the metal must be thoroughly clean, as must the electrodes themselves, and the designated settings for amount of current, duration of current, and electrode pressure must be very accurately maintained. Special offset electrodes have to be made to reach certain jobs and deep-throated machines have been developed for large sheet work.

It is felt that a much wider application will be found for this process as the industry progresses, as it is possible to weld skin sections without leaving any surface obstructions.

CHAPTER 11

THE ASSEMBLY SHOPS

In recent years it has become increasingly evident that the trend is towards "mass production" in the manufacture of aircraft. This is true not only in detail manufacture but also in assembly. Obviously, there would be little gain if each part were made on a high-production basis, and the assembly still were done by the old method of building the airplane from propeller to rudder in one department from a collection of parts.

High-production Assembly—Mass production really begins with the aeronautical engineer. He designs the airplane so that it is possible to build it as a number of major assemblies, which are in turn composed of a greater number of subassemblies. By this method, the efforts of more men are directed to the building of each airplane because they are split up into different departments. Less time is also taken at final assembly, because at this stage a large percentage of the assembly work has already been completed.

Production Department—As can readily be imagined, however, this system would fail entirely unless a very close co-ordination existed between each department. The inability of one department to produce its assembly at a given time would disrupt the whole system. The production department, therefore, is charged with the responsibility of co-ordination. This is a gigantic task, when one considers that 30,000 or 40,000 parts must flow like tributaries throughout the factory to complete one average military airplane. Not only must they flow, but they must be on schedule. The production manager, through his corps of assistants, supplies each and every manufacturing foreman with the necessary blueprints and all the information he requires as to what he should build or make, when it is required, and what material is available. Each foreman is given a production schedule which shows him the production requirements of the entire factory, including that of his own department. It is his duty to use his technical skill to the best advantage, to keep the "flow" of parts or assemblies uninterrupted. As some assemblies require more man hours to complete than others, the production department must see that the toolroom gets busy with those particular tools. This is doubly important, because large assembly fixtures require many toolroom hours to complete.

Fuselage—In many cases the fuselage of the airplane is divided into three portions and spliced together at assembly. This is a very good design, as it lends itself readily to high-production methods and does not retard flying efficiency. The three sections are built on separate assembly lines, each unit moving progressively until it is completed and transported to the splicing fixture.

The nose section, if the design permits, may carry the pilot's cockpit, enabling a vast number of installations to be made without the difficulty of inaccessibility, as would be the case if the fuselage were built in one piece. The center section of the fuselage carries the wing roots. The flap and aileron controls, as well as the landing-gear retracting mechanism, may also be installed in this major assembly. The center-section assembly fixtures are usually the most elaborate of all. The structural members such as bulkheads, frames, longerons and stringers are all located in these fixtures and riveted together.

The front and rear center-section spars, to which are attached the outer wings, are located in place very accurately. Any error in this location would be multiplied at the wing tip and would throw the entire airplane out of symmetry.

The tail section of the fuselage is built similarly, in a

large assembly fixture, but the empennage is not installed until the final assembly. In all three sections, the skin is applied while in the fixtures.

A matter of major importance affecting a fuselage of this type is the absolute necessity of having the splicing holes match up with one another. These holes, of course, must be drilled from bushings in the fixtures, and they must be reamed to size to prevent the bolts from having any play.

Outer Wings—The outer wings are themselves built in another department and are also divided into subassemblies. Here again the use of large assembly fixtures plays a great part in maintaining accuracy. The front and rear spars are assembled as individual units, and it is essential that all the holes and as many attaching parts as possible be added at this stage. This includes all stiffeners and control-pulley brackets. The spars are then placed in another fixture for installation of the wing ribs, which are riveted in place. The ribs are divided into three portions, the nose ribs which form the leading edge of the wing, the center ribs which keep the spars in their proper relationship and form the main portion of the wing, and the tail ribs which form the trailing edge of the wing. Over this entire structure the skin is riveted, using flush-type rivets. The tail surfaces, although much smaller, are also of spar-and-rib construction.

Engine Mount—The only part of the modern airplane in which the steel-tube construction is still used is the engine mount. Considering the light weight of the engine mount, it is astounding that it can withstand the terrific strain that exists between engine and wing. The steel used is 4130, in tube form. These numerals indicate the S.A.E. (Society of Automotive Engineers) specification. The principal alloying element in this steel is molybdenum, which gives it a very high tensile strength, due to its very close grain structure.

The tubes are cut to length and very carefully fitted into place in the welding fixture. Special gussets are also fitted at the main stress points. Oxyacetylene welding is used

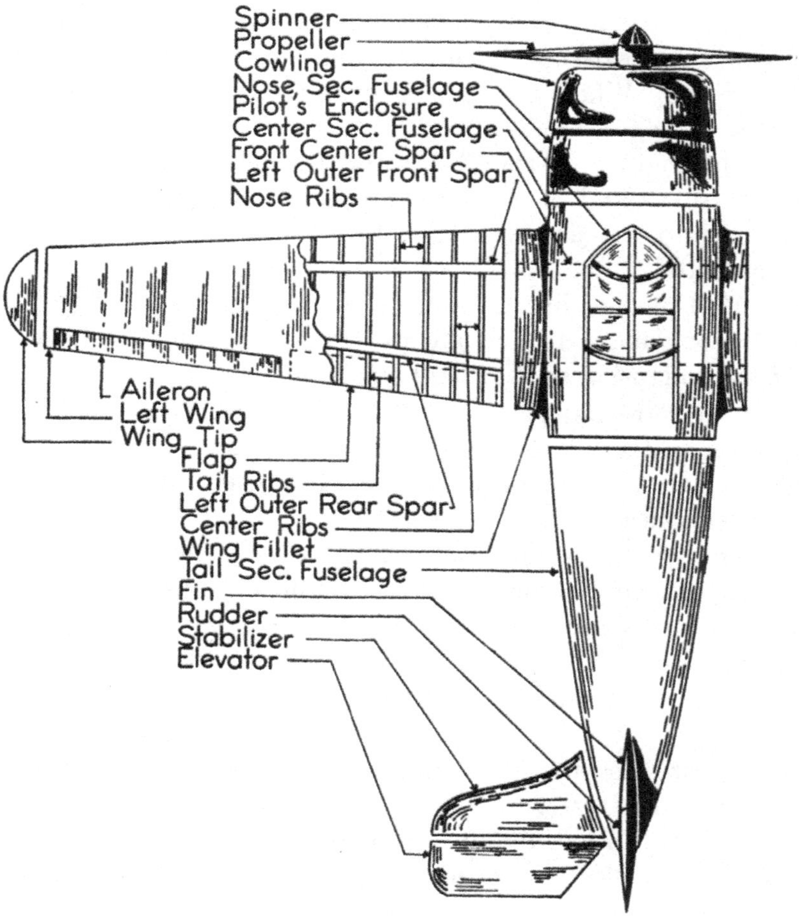

FIG. 48

here, and very skillful welders must be chosen for this work. They must be thoroughly familiar with the shrinkages that affect such welded structures. Every two or three months they must submit sample pieces of their work to be tested for penetration and tensile strength. After welding, the engine mount is checked for distortion and straightened if necessary. It is then jig drilled and reamed for the connecting bolts. Before being put into service, a special liquid is forced through the inside of every tube, which congeals on the inner walls and prevents internal corrosion. Where necessary, a small hole is drilled to admit this liquid.

The engines are installed in the mount and all the necessary tubes and controls are assembled before the engine goes into the airplane. In this way, the work is more accessible.

All tube bending is done by one department. Special tube-bending machines are used which prevent the tubes from collapsing as they are being bent.

An electrical department takes care of wiring all junction boxes, switches and other electrical work that may be necessary. In military airplanes, the various armament-control systems are electrically operated.

The engine cowlings are made by sheet-metal workers. Although very accurate jigs and fixtures are used throughout, a slight amount of handwork is necessary to match the curved sheet-metal parts. This is especially true where engine nacelle and wing fillets are concerned. Such items as bomb bay doors, landing-gear doors, and various access doors are also fabricated in this department, with the use of special fixtures.

It is only when the final-assembly department is reached that the collective efforts of the entire personnel can be visualized or appreciated. The ease with which the major assemblies, already mentioned, go together is tribute to the efficiency of the aircraft industry. In this department all the radio equipment is installed; engine controls and connections are completed; flap and landing-gear mechanisms are hooked up and tested; and all other surface controls are connected and adjusted. The entire array of navigation instruments is also installed.

It is well for the apprentice to realize that in any of the assembly departments just mentioned there is ample scope for initiative. The methods and sequences of today do not have to be followed tomorrow, if someone with foresight and ingenuity presents a better alternative. When the completed airplane flies over the factory with a thunderous roar on its test flight, there is not a man in the organization that does not thrill to the thought that he contributed to the flight of that noisy but efficient mechanical bird.

Assembly Apprentice's Tool Kit—The starting tools that the assembly apprentice will require are as follows:

1. Ball-peen hammer (8 oz.)
2. 6-in. flexible scale
3. 12-in. combination square with protractor
4. Center punch
5. Screwdrivers (1 large, 1 small)
6. Diagonal-side cutting pliers
7. Square-nose pliers
8. Open-end wrenches (1 set)
9. Socket wrenches (1 set)
10. Adjustable wrenches (1 large, 1 small)
11. Six-foot flexible rule

Assembly Tools—The principal machines that the apprentice will come in contact with in assembly work are:

1. Drill press
2. High-speed riveting hammer
3. Rivet squeezer
4. One-shot impact hammer
5. Hand electric drill

CHAPTER 12

GETTING AHEAD

It is hoped that the foregoing chapters have given the schoolboy some idea of the workings within the aircraft factory, so that he may now be in a better position to decide his future course.

It is the natural ambition of the average boy to get ahead, but in order to do so he must not let the grass grow under his feet. It is not enough for him to get a job in some aircraft factory, and then let things take their own course. Getting the job is really only the start. Besides studying his job in all its phases, he must think about changing his mental attitude from that of the schoolboy to that of the prospective leader of men.

He may have thought when he left school that his education was complete, but it stuns him to find out how little he knows. At school he was led along the paths of learning by his teachers, but in the factory he must be more independent. After preliminary instruction, he must develop his own initiative, he must ask questions rather than wait to be told.

Everything in the factory is different for him, but not everything is pleasantly different. He misses the schoolmates who surrounded him and who looked to him for leadership during that last year of school. New friends are not easy to make either, for a boy at that sensitive age.

Although the school bell has been replaced by the factory buzzer, the apprentice must realize that he is still learning, but instead of struggling with the three R's, he is now confronted with the three M's of industry—men, machines and materials.

Industry is the life blood of the nation, and the three M's are its dominating factors. The future position of the apprentice must therefore rise or fall in direct proportion to his knowledge of men, machines and materials. While there is no such thing as a complete knowledge of these, it is to his advantage to acquire as much of it as possible.

The rut in which many get stuck is in believing that the employee is merely a cog in the industrial wheel, being driven to "mesh in" momentarily, somewhere in the circle.

Men—In reality, men are the driving power of industry. They think, they act, they create ideas, they build; often they are defeated, but they rise again stronger than ever. They also possess a "divine individualism" that defies mechanical reproduction and which stamps every man as being distinct from his neighbor. Every cog in the wheel, it will be recalled, is alike, and each one travels in the same circle, retracing and coming back to where it started. In contrast, the influence of man spreads onward, outward and upward, and you, the apprentice, are entering the ranks of industry, with the inherent desire of carrying the torch of civilization forward with you. How far shall you have carried it when the time comes for you to relinquish it to another? That will be the measure of your success.

There is no room for one-track thinking in industry, and it is man's task so to control and divert properly the varieties of character, temperament and intellect that the efforts of all shall be directed to the common good.

Every source of mechanical power known to man is usable only to the extent to which it can be controlled. The control of men who are both the intellectual driving power and guiding force of industry presents many complicated problems, usually termed "personnel problems." Much has been written in recent years regarding the responsibilities of industry in this direction, but very little about the responsibilities that must be carried by the individual. After all, the whole personnel problem is distinctly two-sided. Industry as a whole has risen to its part of the bargain magnificently, with the result that working and living conditions are more favorable to the worker than at any other period in the history of the world.

Most of the unhappiness that attends us is due to individual jealousy, greed and lust for power, so every apprentice should take steps to control these vices within himself. The forces that motivate all men are very much alike, but the courses followed are vastly different. For instance, some men are jealous, others impartially aggressive. Some men are greedy, others thrifty. Some have a lust for power, others have a natural ambition. It is only when the latter is tinged with lust that it becomes one of the baser instincts.

The whole aspect of human behavior in industry is worthy of a more versatile pen than mine, but certainly, if a man strives to keep his body healthy, his character irreproachable, and his mind active and clean, he stands a more than even chance of attaining his natural ambitions.

Machines—So much for the first of the three M's but how about the machines of industry? They must be considered in second place, because they are entirely subordinate to the will of man, yet must take priority over materials because they are a higher development of the materials themselves. The most important machines, as we have seen, are those classified as machine tools, because, not only are they responsible for the high production of all the metallic necessities of life, but also they produce the parts from which are assembled all other machines. It is interesting to note that the machine tool also reproduces itself.

The subsidiary machines of industry, from the simple egg-beater of the housewife to the thundering locomotive, have become so essential a part of our daily lives that this has very aptly been termed the Machine Age. It is for the apprentice to remember, however, that in taking second place only to man himself, the machine demands the care and upkeep without which delicate mechanisms cannot survive.

Materials—The materials of industry are the links which connect our artificial life with Mother Nature herself. Whether mined from the earth or grown in the forest, they represent a tremendous amount of human enterprise.

In the case of the steels and alloys which interest us in the aircraft industry, it would seem that the metallurgists have produced a metal for every requirement. Not only is it possible to obtain extreme hardness, toughness, tensile strength or lightness in metals, but combinations of these properties have been developed by modern metallurgy. This widens the scope of the aero engineer immeasurably. Do you who have witnessed modern aircraft in fabrication and seen it in flight, with its engines beating out a furious but rhythmic symphony of power, ever think of the unsung heroes who toil in the background, perfecting the materials of this industry?

The steel that makes the machine tool groan, the aluminum that is beaten and twisted into shape are not merely the inanimate substances that one may suppose. They have emerged from the furnaces of their re-creation with a definite purpose to fulfill in this world. How they fulfill it depends largely upon how you, the men of industry, treat or mistreat them.

Yes, the three M's of industry, properly applied, form the basis of our present civilization, and a very large part of the responsibility in maintaining this rests with you, the apprentice.

SELF-ANALYSIS FOR JOB SELECTION

CHART NO. I

DETERMINING FACTORS FOR SHOP OR ENGINEERING

Ask yourself these questions.	*If So, Check*	
Do you like Mechanical Drawing?	( )	If most of your check marks are in this group, you are suited for Engineering work, in which case you should disregard Chart No. 2.
Do you like higher Mathematics?	( )	
Are you more interested in theory than in practice?	( )	
Have you ever designed your own model aircraft?	( )	
Does the study of Aerodynamics interest you?	( )	
Do you like to read technical books?	( )	
Are you prepared to continue your studies intensively?	( )	
Can you write a good report?	( )	
Are you more interested in why they fly than in how they are built?	( )	
Would you consider drawing board work too confining?	( )	

	If So, Check	
Are you mechanically inclined?	( )	If most of your check marks are in this group, you are suited for shop work. Now fill in Chart No. 2.
Do you like manual labor?	( )	
Do you like blueprint reading?	( )	
Do you like working around machinery?	( )	
Do you mind shop-soiled hands?	( )	
Do you like working with metal?	( )	
Do you have any mechanical hobbies?	( )	
Do you experience pride upon completion of a mechanical job?	( )	
Are you handy with tools?	( )	
Do you like Geometry?	( )	

CHART NO. II

To Determine Which Shop Department You Are Suited For

Ask yourself these questions.	*If So, Check*	
Do you like working with precision tools?	( )	If most of your check marks are in this group, you are suited for the Tool Room.
Are you inventive?	( )	
Are you interested in high production?	( )	
Do you have an eye for accuracy?	( )	
Does the heat treatment of steel interest you?	( )	
Do you like Plane Geometry?	( )	If most of your check marks are in this group, you are suited for Sheet Metal.
Do you feel an interest in sheet-metal layout?	( )	
Are you clever with your hands?	( )	
Are you artistic?	( )	
Would you be interested in punch-press operation?	( )	
Do you realize the importance of machine tools?	( )	If most of your check marks are in this group, you are suited for the Machine Shop.
Do you like repetitive work?	( )	
Do cutting tools interest you?	( )	
Does an intricate machine interest you?	( )	
Are you willing to study the five basic methods of machining?	( )	
Do you prefer working at a bench?	( )	If most of your check marks are in this group you are suited for Assembly Departments.
Are you interested in working on the airplane itself?	( )	
Are you prepared to work from scaffolding?	( )	
Would you like working with air-operated riveting tools?	( )	
Would you be interested in making assemblies from parts made by other departments?	( )	

INDEX

The Aviation Collection by Sportsman's Vintage Press

www.SportsmansVintagePress.com

Aircraft Construction Handbook	by Thomas A. Dickinson
Aircraft Sheet Metal Work	by C. A. LeMaster
The Aircraft Apprentice	by Leslie MacGregor
Aircraft Woodwork	by Col. R. H. Drake
Aircraft Welding	by Col. R. H. Drake
Aircraft Sheet Metal	by Col. R. H. Drake
Aircraft Engines	by Col. R. H. Drake
Aircraft Electrical and Hydraulic Systems, and Aircraft Instruments	by Col. R. H. Drake
Aircraft Engine Maintenance and Service	by Col. R. H. Drake
Aircraft Maintenance and Service	by Col. R. H. Drake

Made in the USA
Monee, IL
04 November 2026

67387322R00085